BrightRED Study Guide

PHYSICS

Andrew McGuigan

First published in 2015 by:
Bright Red Publishing Ltd
1 Torphichen Street
Edinburgh
EH3 8HX

MIX
Paper from
responsible sources
FSC® C013254
FSC
www.fsc.org

A CIP record for this book is available from the British Library.

ISBN 978-1-906736-73-6

With thanks to:
PDQ Digital Media Solutions Ltd, Bungay (layout) and Project One Publishing Solutions (copy-edit).
Cover design and series book design by Caleb Rutherford – e i d e t i c.

Acknowledgements
Every effort has been made to seek all copyright-holders. If any have been overlooked, then Bright Red Publishing will be delighted to make the necessary arrangements.

Permission has been sought from all relevant copyright holders and Bright Red Publishing are grateful for the use of the following:

Caleb Rutherford e i d e t i c (p 10); Caleb Rutherford e i d e t i c (p 15); Image licensed by Ingram Image (p 21); stu smith (CC BY-ND 2.0)[1] (p 22); NASA/JPL-Université Paris Diderot – Institut de Physique du Globe de Paris (p 23); shirokazan (CC BY 2.0)[2] (p 23); NASA/JPL (p 29); NASA/JPL/Space Science Institute (p 29); NASA, ESA, A. Bolton (Harvard-Smithsonian CfA) and the SLACS Team (p 35); ESA/NASA (p 36); koya79/ iStock.com (pp 36 & 37); AllenMcC. (CC BY-SA 3.0)[3] (p 36); Image licensed by Ingram Image (p 44); NASA/ SDO/AIA/HMI/Goddard Space Flight Center (p 47); NASA, ESA, J. English (U. Manitoba), and the Hubble Heritage Team (STScI/AURA) (p 49); NASA, ESA, H. Bond (STScI) and M. Barstow (University of Leicester) (p 51); NASA, NOAO, ESA, the Hubble Helix Nebula Team, M. Meixner (STScI), and T.A. Rector (NRAO) (p 55); NASA/CXC/IAFE/G. Dubner et al & ESA/XMM-Newton (p 57); Serendipity Diamonds (CC BY-ND 2.0)[1] (p 57); CERN Geneva (p 67); Joshua Strang (p 69); Ron Horii (p 89); dun_deagh (CC BY-SA 2.0)[4] (p 115); Image licensed by Ingram Image (p 12); Image licensed by Ingram Image (p 126).

Exam question taken from 2002 Advanced Higher Physics paper © Scottish Qualifications Authority (n.b. solutions do not emanate from the SQA) (p 6)

(CC BY-ND 2.0)[1]   http://creativecommons.org/licenses/by-nd/2.0/
(CC BY 2.0)[2]   http://creativecommons.org/licenses/by/2.0/
(CC BY-SA 3.0)[3]   http://creativecommons.org/licenses/by-sa/3.0/
(CC BY-SA 2.0)[4]   https://creativecommons.org/licenses/by-sa/2.0/

Printed and bound in the UK by Martins the Printers.

CONTENTS

INTRODUCTION

INTRODUCING CfE ADVANCED HIGHER PHYSICS

AIMS AND STRUCTURE OF THE BOOK

This book covers the syllabus for the CfE Advanced Higher Physics course and should complement the coursework done in class. The book will help students succeed in the final exam by presenting the subject arrangements in an attractive and concise format.

Each sub-topic in the Specifications is presented in a double-page spread making the book ideal both for revision and self-study.

Each double-page spread covers the physics content in a logical and accessible way, and makes full use of graphics and colour illustrations to support your learning. Spreads contain the mandatory course physics information and **Worked examples** demonstrating how key physics relationships are used to solve numerical problems. There are also **Exercises** for you to try with answers available on the **Bright Red Digital Zone**. Examinable derivations are included in the text where appropriate.

On each spread you will find **Don't Forget** tips which highlight key points or common mistakes. **Internet Links** suggest sites for additional information, useful videos, interactive simulations or practical applications of the topic in question. Each double-page spread finishes with a **Things to do and think about** section designed to extend and expand your knowledge and interest in physics and its applications.

The treatment of uncertainties at Advanced Higher level is covered in three double-page spreads. This treatment can apply to all units in the course. There is also advice on the production of the **project report** and there are appendices with guidance on open-ended questions and on understanding accuracy and precision in physics.

COURSE STRUCTURE AND ASSESSMENT

The AH Physics course comprises these units:

- Rotational Motion and Astrophysics
- Quanta and Waves
- Electromagnetism (½ unit)
- Investigating Physics (including the production of a project report) (½ unit).

Students are required to pass an internal assessment on completion of each unit.

The external assessment consists of two parts.

- A written examination paper of 2½ hours duration with an allocation of 140 marks, which will be scaled to 100. In addition to the questions based around the mandatory coursework there will be two open-ended questions and at least one question using given knowledge which is not in the syllabus. You will be provided with a Data Sheet containing relevant data, and a Relationships Sheet containing formulae and a periodic table.

- A project report on the investigation will be marked externally by an SQA marker with a total allocation of 30 marks for the report.

The total for the two external assessments is 130 marks. The course award is graded A, B, C or D depending on the overall mark out of 130.

RELATIONSHIPS REQUIRED FOR ADVANCED HIGHER PHYSICS

$v = \frac{ds}{dt}$

$a = \frac{dv}{dt} = \frac{d^2s}{dt^2}$

$v = u + at$

$s = ut + \frac{1}{2}at^2$

$v^2 = u^2 + 2as$

$\omega = \frac{d\theta}{dt}$

$\alpha = \frac{d\omega}{dt} = \frac{d^2\theta}{dt^2}$

$\omega = \omega_0 + \alpha t$

$\theta = \omega_0 t + \frac{1}{2}\alpha t^2$

$\omega^2 = \omega_0^2 + 2\alpha\theta$

$s = r\theta$

$v = r\omega$

$a_t = r\alpha$

$a_r = \frac{v^2}{r} = r\omega^2$

$F = \frac{mv^2}{r} = mr\omega^2$

$T = Fr$

$T = I\alpha$

$L = mvr = mr^2\omega$

$L = I\omega$

$E_K = \frac{1}{2}I\omega^2$

$F = G\frac{Mm}{r^2}$

$V = -\frac{GM}{r}$

$v = \sqrt{\frac{2GM}{r}}$

apparent brightness, $b = \frac{L}{4\pi r^2}$

power per unit area $= \sigma T^4$

$L = 4\pi r^2 \sigma T^4$

$r_{Schwarzschild} = \frac{2GM}{c^2}$

$E = hf$

$\lambda = \frac{h}{p}$

$mvr = \frac{nh}{2\pi}$

$\Delta x \Delta p_x \geqslant \frac{h}{4\pi}$

$\Delta E \Delta t \geqslant \frac{h}{4\pi}$

$F = qvB$

$\omega = 2\pi f$

$a = \frac{d^2y}{dt^2} = -\omega^2 y, \; y = A\cos\omega t$
$\qquad \text{or } y = A\sin\omega t$

$v = \pm\omega\sqrt{(A^2 - y^2)}$

$E_K = \frac{1}{2}m\omega^2(A^2 - y^2)$

$E_P = \frac{1}{2}m\omega^2 y^2$

$y = A\sin 2\pi\left(ft - \frac{x}{\lambda}\right)$

$\Phi = \frac{2\pi x}{\lambda}$

optical path difference $= m\lambda$
or $\left(m + \frac{1}{2}\right)\lambda$ where $m = 0, 1, 2...$

$\Delta x = \frac{\lambda l}{2d}$

$d = \frac{\lambda}{4n}$

$\Delta x = \frac{\lambda D}{d}$

$n = \tan i_P$

$F = \frac{Q_1 Q_2}{4\pi\varepsilon_0 r^2}$

$E = \frac{Q}{4\pi\varepsilon_0 r^2}$

$V = \frac{Q}{4\pi\varepsilon_0 r^2}$

$F = QE$

$V = Ed$

$F = IlB\sin\theta$

$B = \frac{\mu_0 I}{2\pi r}$

$c = \frac{1}{\sqrt{\varepsilon_0\mu_0}}$

$t = RC$

$X_C = \frac{V}{I}$

$X_C = \frac{1}{2\pi f C}$

$\varepsilon = -L\frac{dI}{dt}$

$E = \frac{1}{2}LI^2$

$X_L = \frac{V}{I}$

$X_L = 2\pi f L$

$\frac{\Delta W}{W} = \sqrt{\left[\frac{\Delta X}{X}\right]^2 + \left[\frac{\Delta Y}{Y}\right]^2 + \left[\frac{\Delta Z}{Z}\right]^2}$

$\Delta W = \sqrt{\Delta X^2 + \Delta Y^2 + \Delta Z^2}$

KINEMATIC RELATIONSHIPS

In earlier studies of physics, we used three equations of motion for objects moving in a straight line with constant acceleration.

- $v = u + at$
- $s = ut + \frac{1}{2}at^2$
- $v^2 = u^2 + 2as$

where the symbols have their usual meanings.

In Advanced Higher Physics, we derive these equations using calculus.

CALCULUS METHODS TO DERIVE EQUATIONS OF MOTION

Velocity v is defined as the rate of change of displacement s or $v = \frac{ds}{dt}$

Acceleration a is defined as the rate of change of velocity v or $a = \frac{dv}{dt}$

Combining, we get $a = \frac{dv}{dt} = \frac{d}{dt}\left(\frac{ds}{dt}\right) = \frac{d^2s}{dt^2}$

Deriving $v = u + at$

$$\frac{dv}{dt} = a$$
$$dv = adt$$
$$\int_u^v dv = \int_0^t adt$$
$$[v]_u^v = [at]_0^t$$
$$v - u = at$$
$$v = u + at$$

Alternative method

$$\int dv = \int adt$$
$$v = at + c$$

at time $t = 0$, $v = u$

so $u = 0 + c$ and $c = u$

$$v = u + at$$

Deriving $s = ut + \frac{1}{2}at^2$

$$\frac{ds}{dt} = v$$
$$ds = vdt$$
$$\int_0^s ds = \int_0^t vdt = \int_0^t (u + at)dt = \int_0^t udt + atdt$$
$$[s]_0^s = [ut + \frac{1}{2}at^2]_0^t$$
$$s = ut + \frac{1}{2}at^2$$

Alternative method

$$\int ds = \int vdt = \int(u + at)dt$$
$$s = ut + \frac{1}{2}at^2 + c$$

at time $t = 0$ $s = 0$ so $c = 0$

$$s = ut + \frac{1}{2}at^2$$

Deriving $v^2 = u^2 + 2as$

Having derived $v = u + at$ and $s = ut + \frac{1}{2}at^2$, using calculus rearranging $v = u + at$ to get $t = \frac{v - u}{a}$

Substitute this into $s = ut + \frac{1}{2}at^2$

$$s = u\frac{(v - u)}{a} + \frac{1}{2}a\frac{(v - u)^2}{a^2}$$
$$2as = 2uv - 2u^2 + v^2 - 2uv + u^2$$
$$v^2 = u^2 + 2as$$

DON'T FORGET ⊕

The order and detail of the various steps of the derivations are important.

Example

An arrow is fired vertically into the air, and its vertical displacement s is given by

$s = 44 \cdot 1t - 4 \cdot 9t^2$ metres, where t is in seconds

a Find an expression for the velocity of the arrow.

b Find the acceleration of the arrow.

c Calculate the initial velocity of the arrow.

d Calculate the maximum height reached by the arrow.

Solution:

a $v = \frac{ds}{dt} = 44 \cdot 1 - 9 \cdot 8t$

b $a = \frac{dv}{dt} = -9 \cdot 8 \, ms^{-2}$

c $v = 44 \cdot 1 - 9 \cdot 8t$ when $t = 0$
$= 44 \cdot 1 - 0$
$v = 44 \cdot 1 \, ms^{-1}$

d Max height when $v = 0$:
$44 \cdot 1 - 9 \cdot 8t = 0$, $t = \frac{44 \cdot 1}{9 \cdot 8} = 4 \cdot 5 \, s$
$s = 44 \cdot 1 \times 4 \cdot 5 - 4 \cdot 9 \times 4 \cdot 5^2 = 99 \cdot 2 \, m$

contd

Displacement-time graphs

The gradient of a displacement–time graph of an object gives information about the velocity of the object. The following graphs have different gradients each representing a different motion.

blue line: constant gradient = **constant velocity**
red line: $\frac{ds}{dt}$ is **increasing**.
This graph represents **acceleration**.
orange line: $\frac{ds}{dt}$ is **decreasing** representing **deceleration** (or negative acceleration).

The gradient at any point on the red or orange lines represents the **instantaneous velocity** at that point.

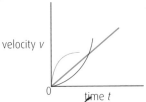

Velocity-time graphs

The gradient of a velocity–time graph of an object gives information about the acceleration of the object. The following graphs have different gradients each representing a different motion.

blue line: constant gradient = **constant acceleration**
red line: $\frac{ds}{dt}$ (or $\frac{d^2s}{dt^2}$) is **increasing**.
This graph represents **increasing acceleration**.
orange line: $\frac{ds}{dt}$ (or $\frac{d^2s}{dt^2}$) is **decreasing**.
This graph represents **decreasing acceleration**.

The gradient at any point on the red or orange lines represents the **instantaneous acceleration** at that point.

Area under velocity-time graph

The displacement of an object can be found by calculating the area under a velocity–time graph. It may be necessary to use integration.

Example

A car takes 12 seconds to travel between two sets of traffic lights. The graph shows the velocity of the car as it makes the journey.

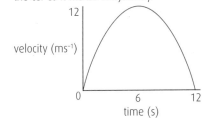

The velocity of the car v is given by the expression $v = 3t - 0\cdot25t^2$ where v is in ms^{-1} and t is in seconds. Show that the distance travelled between the two traffic lights is 72 m.

VIDEO LINK

Check out the clip at www.brightredbooks.net to learn more about deriving kinematic relationships using calculus.

Solution:

distance = area under the graph between zero and 12 seconds.

$s = \int_0^{12}(3t - 0\cdot25t^2)dt$

$= \left[\frac{3t^2}{2} - \frac{0\cdot25t^3}{3}\right]_0^{12}$

$= \left[\frac{3 \times 12^2}{2} - \frac{0\cdot25 \times 12^3}{3}\right] - [0]$

$= 216 - 144$

$= 72\,m$

Alternative method

$s = \int(3t - 0\cdot25t^2)dt$

$= \frac{3}{2}t^2 - \frac{1}{12}t^3 + c$

at $t = 0$, $v = 0$ so $c = 0$

$s = \frac{3}{2} \times 12^2 - \frac{1}{12} \times 123 - 0$

$= 216 - 144 - 0$

$= 72\,m$

DON'T FORGET

The first line must be correct in a 'show' question and the term dt must be included.

THINGS TO DO AND THINK ABOUT

The displacement s of an object is given by $s = 20t - t^3$ after starting from rest at time $t = 0$.

a Find the speed and acceleration of the object at time $t = 3$ s.

b Calculate the time when the object is stationary.

ONLINE TEST

Want to test yourself on kinematic relationships? Head to www.brightredbooks.net

ANGULAR MOTION

You have studied linear motion in National 5 and Higher Physics. In Advanced Higher Physics, this work is extended to cover **circular motion** where objects move along circular paths rather than straight-line paths.

RADIAN MEASURE

An angle in **radians** is defined as $\theta = \frac{arc\ length}{radius} = \frac{s}{r}$.

This relationship is given in the physics relationships sheet as $\mathbf{s = r\theta}$.

The angle θ is known as the **angular displacement**.

ANGULAR VELOCITY $\omega = \frac{\theta}{t}$

In circular motion, **angular velocity** ω is the measure of the **angle swept per second**.

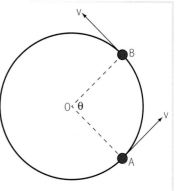

Consider an object moving at a steady linear speed v in a circle with centre O. The angle θ is swept out by the radius as the object moves from A to B in a time of t seconds.

θ is measured in **radians**, so ω is measured in **radians per second** or **rads^{-1}**.

Example

A wind generator makes 6 complete revolutions in a time of 10 seconds. Calculate the angular velocity of the wind generator.

Solution:

1 complete revolution = 2π radians

$\omega = \frac{\theta}{t} = \frac{6 \times 2\pi}{10} = 3 \cdot 8$ rads^{-1}

ω can also be calculated using the frequency of rotation or by differentiation.

The time for one complete revolution is called the **period, T**.

$\omega = \frac{\theta}{t} = \frac{2\pi}{T}$

The frequency $\mathbf{f}$ of rotation is related to the period T by $f = \frac{1}{T}$, so substituting gives

$\omega = 2\pi f$

Differentiation can be used to calculate ω if an expression for θ as a function of time is given.

$\omega = \frac{d\theta}{dt}$

⚙ EXERCISE

1   A microwave tray rotates with a period of 4 seconds. Calculate its angular velocity.

2   Calculate the angular velocity of the Earth as it rotates about its polar axis.

3   Mars rotates about its polar axis with an angular velocity of $7 \cdot 1 \times 10^{-5}$ rads^{-1} and rotates around the Sun with an angular velocity of $1 \cdot 06 \times 10^{-7}$ rads^{-1}. Calculate the length of a
   a   Martian day
   b   Martian year.

4   How many radians does the Earth turn through in 1 year?

ANGULAR VELOCITY AND TANGENTIAL SPEED

Objects moving in a circle have angular velocity and a **linear speed in the direction of the tangent**. This **linear** or **tangential speed v** is related to the **angular speed** ω.

Consider one complete orbit in a circle of **radius r**. The period of rotation is T.

$$v = \frac{distance}{time} = \frac{circumference}{time} = \frac{2\pi r}{T} = \frac{2\pi}{T}r = \omega r$$

$$v = r\omega$$

Example

A ceiling fan rotates at 150 rpm. Each blade of the fan is 45 cm long.

a Calculate the angular velocity of the fan.

b Calculate the speed of the tip of each blade.

c Calculate the speed of a point halfway along a blade.

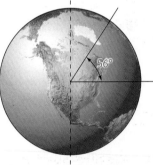

Solution:

a $\omega = \frac{\theta}{t} = \frac{150 \times 2\pi}{60} = 15{\cdot}7$ rads^{-1}

b $v = \omega r = 15{\cdot}7 \times 0{\cdot}45 = 7{\cdot}1$ ms^{-1}

c $v = \omega r = 15{\cdot}7 \times 0{\cdot}225 = 3{\cdot}5$ ms^{-1}

Note that **points closer to the axis** of rotation will have **lower tangential speeds** than points further from the axis. The r in $v = \omega r$ is the **distance from the axis of rotation** to the point whose tangential speed is being calculated.

⚙ EXERCISE

5 Assuming the Earth to be spherical with radius $6{\cdot}4 \times 10^6$ m, calculate

  a the tangential speed of a person standing on the Equator.

  b the tangential speed of a person standing in central Scotland, which has a latitude of 56°. The latitude of a point on the Earth's surface is the angle from the equator (see diagram).

6 Calculate the linear speed of the tip of the minute hand of a watch. The minute hand is 1·2 cm long and moves continuously.

7 Calculate the linear speed of the Earth as it orbits the Sun. The Earth's mean orbit radius is $1{\cdot}5 \times 10^{11}$ m.

ONLINE

Visit www.brightredbooks.net for more information and examples of angular velocity

💭 THINGS TO DO AND THINK ABOUT

The Advanced Higher Physics course will test your data handling abilities in unfamiliar situations. Try the following more complex problem involving circular motion.

A star is 3000 light years from the centre of its rotating constellation and moves with a linear speed of 113 km s^{-1}. Show that it makes one complete orbit every 50 million years.

ONLINE TEST

Head to www.brightredbooks.net for a test on angular motion.

ANGULAR ACCELERATION

ANGULAR ACCELERATION

When the **angular velocity of a rotating object changes**, it is said to have an **angular acceleration**. Angular acceleration has the symbol α and unit **rads^{-2}**. A constant angular acceleration of 3 rads^{-2} means the angular velocity increases by 3 rads^{-1} every second.

Angular acceleration can be described mathematically.

$\alpha = \frac{d\omega}{dt}$ angular acceleration is the rate of change of angular velocity.

$\alpha = \frac{d^2\theta}{dt^2}$ since $\omega = \frac{d\theta}{dt}$ then α is the second derivative of $\theta(t)$.

These two formulae will give the **instantaneous angular acceleration** at time **t**.

Example

The angular displacement of a particle performing circular motion is given by

$\theta = 4t^2 + 3t + 2$ radians.

Calculate the angular velocity and the angular acceleration of the particle after 5 s.

Solution:

$\omega = \frac{d\theta}{dt} = 8t + 3 = (8 \times 5) + 3 = 43$ rads^{-1}

$\alpha = \frac{d\omega}{dt}$

$= 8$ rads^{-2} (= a constant acceleration independent of time t)

⚙ EXERCISE

1 The angular displacements of two objects A and B are given by

object A: $\theta = -2 - 0\cdot6t + 0\cdot35t^2$ object B: $\theta = 0\cdot5t + 0\cdot15t^3$

Calculate the angular acceleration of each object after 4 seconds.

CONSTANT ANGULAR ACCELERATION

If the angular acceleration is **constant**, then the following expression can be used to calculate α.

$\alpha = \frac{\omega - \omega_0}{t}$ where ω_0 = initial angular velocity (rads^{-1})

ω = final angular velocity (rads^{-1}) after time t (s)

Example

During take-off, the main rotor blades of a helicopter increase their rate of rotation uniformly from rest to 450 rpm in 8 seconds. Calculate the angular acceleration of the rotor blades.

Solution:

$450 \text{ rpm} = \frac{450 \times 2 \times 3\cdot14}{60}$

$= 47\cdot1$ rads^{-1}

$\alpha = \frac{\omega - \omega_0}{t}$

$= \frac{47\cdot1 - 0}{8}$

$= 5\cdot9$ rads^{-2}

Note that the tip of the blade will have an increasing **linear** speed as the angular speed increases, giving rise to the concept of **tangential acceleration** a_t.

Tangential acceleration can be thought of as an instantaneous linear acceleration at a point on the circle.

If the helicopter blade is 6·5 m long, then the tangential acceleration of the tip of the blade can be calculated

$a_t = \alpha r = 5\cdot9 \times 6\cdot5 = 38 \text{ ms}^{-2}$.

DON'T FORGET ➕

Tangential acceleration is a linear acceleration with unit ms^{-2} and is different from angular acceleration with unit rads^{-2}.

KINEMATIC RELATIONSHIPS FOR CONSTANT ANGULAR ACCELERATION

The following equations for constant angular acceleration are exactly analogous to those for constant linear acceleration which were derived on page 4.

$\omega = \omega_0 + \alpha t$ where ω_0 = initial angular velocity (rads^{-1})

$\theta = \omega_0 t + \frac{1}{2}\alpha t^2$ ω = final angular velocity (rads^{-1})

$\omega^2 = \omega_0^2 + 2\alpha\theta$ α = constant angular acceleration (rads^{-2})

θ = angular displacement (rad)

t = time taken (s)

ONLINE

Visit www.brightredbooks.net for some numerical examples on angular acceleration.

Example

A bicycle wheel rotating at 4·20 revolutions per second comes to rest uniformly in a time of 125 seconds.

a Calculate the angular acceleration of the wheel during this time.

b How many revolutions does the wheel make as it decelerates?

Solution:

a $\omega_0 = \frac{\theta}{t} = \frac{4\cdot20 \times 2 \times 3\cdot14}{1} = 26\cdot4$ rads^{-1}.

$\alpha = \frac{\omega - \omega_0}{t} = \frac{0 - 26\cdot4}{125} = -0\cdot211$ rads^{-2}.

b $\theta = \omega_0 t + \frac{1}{2}\alpha t^2$

$= 26\cdot4 \times 125 + (0\cdot5 \times (-0\cdot211) \times 125^2)$

$= 1652$ rad $= \frac{1649}{2 \times 3\cdot14} = 263$ revolutions.

EXERCISE

2   A spin dryer rotating at 800 rpm makes 95 complete revolutions as it slows down uniformly and stops. Calculate the time taken by the spin dryer to decelerate and stop.

3   A spinning disc accelerates uniformly from 30 rpm to 72 rpm in a time of 4 seconds. Calculate the number of revolutions made by the disc in this time.

4   A lawnmower blade rotating at 300 rpm is switched off. It comes to rest after making 4 complete revolutions. Calculate the time taken by the blade to stop.

5   The graph shows how the angular velocity of a rotating disc varies with time.

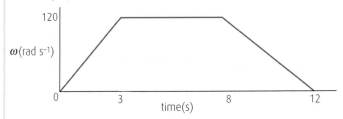

    a   Calculate the initial angular acceleration and final angular acceleration of the disc.

    b   Calculate the total angular displacement of the disc.

    c   How many revolutions does the disc make?

DON'T FORGET

All the data in this example is given to 3 significant figures. The final answer should also have 3 significant figures. Intermediate calculated values should have more than 3 significant figures to avoid incorrect rounding. SQA markers will accept two more or one less significant figures than the correct value in final numerical answers.

THINGS TO DO AND THINK ABOUT

The equations of circular motion are very similar to the equations of linear motion. This is useful as the skills developed using $s = ut + \frac{1}{2}at^2$ are used again with $\theta = \omega_0 t + \frac{1}{2}\alpha t^2$.

However, be careful that you don't accidentally write down something like $s = \omega_0 t + \frac{1}{2}\alpha t^2$.

While this could be considered by many as just a slip, it could be treated as wrong physics in an exam with loss of marks.

ONLINE TEST

Head to www.brightredbooks.net to test your knowledge of angular acceleration.

CENTRIPETAL ACCELERATION AND CENTRIPETAL FORCE

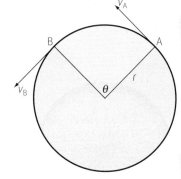

CENTRIPETAL ACCELERATION

When an object moves in a circle of radius r at a steady speed v, its direction is continually changing. Velocity is a vector requiring both magnitude and direction, so the velocity of the object will be changing even if the speed is constant. The object must be accelerating as its velocity is changing. The acceleration, a, is towards the centre of the circle. This acceleration towards the centre of rotation is called the **centripetal acceleration** or **radial acceleration**.

DERIVATION OF $a = \frac{v^2}{r} = \omega^2 r$

The change of velocity from A to B is found using a vector addition diagram.

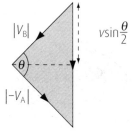

Notice the change in velocity from A to B is directed (down) towards the centre of the circle.

The time to travel from A to B $= \dfrac{arc\ length\ AB}{v} = \dfrac{r\theta}{v}$

acceleration towards centre $= \dfrac{\Delta v}{\Delta t} = \dfrac{2v \sin \frac{\theta}{2}}{\frac{r\theta}{v}}$ (note: $\sin x = x$ for small values of x.)

$$= \dfrac{2v\frac{\theta}{2}}{\frac{r\theta}{v}}$$

$$= \dfrac{v^2}{r} = \dfrac{\omega^2 r^2}{r}$$

$$a = \omega^2 r.$$

Example

A radio-controlled model aircraft moves in a circle of radius 7·5 m. Calculate the centripetal acceleration of the aircraft when

a it travels at a steady speed of 8·2 ms^{-1}

b the aircraft takes 4·5 seconds to make 1 complete circuit.

Solution:

a $a = \dfrac{v^2}{r} = \dfrac{8 \cdot 2^2}{7 \cdot 5} = 9\ \text{ms}^{-2}$

b $\omega = \dfrac{2\pi}{t} = \dfrac{2 \times 3 \cdot 14}{4 \cdot 5} = 1 \cdot 4\ \text{rads}^{-1}$

$a = \omega^2 r = 1 \cdot 4^2 \times 7 \cdot 5 = 14 \cdot 7\ \text{ms}^{-2}$

We have now met three separate accelerations associated with circular motion, and it is important to understand their differences.

Angular acceleration $\alpha = \dfrac{\omega - \omega_0}{t}$ — unit is rads^{-2} increasing/decreasing ω (and v)

Tangential acceleration $a_t = \alpha r$ — unit is ms^{-2} increasing/decreasing v (and ω)

Centripetal acceleration $a = \dfrac{v^2}{r} = \omega^2 r$ — unit is ms^{-2} ω and v constant (for constant a)

⚙ EXERCISE

1 An object moves in a circle with radius 4·1 m and an angular velocity 2·6 rads^{-1}.
The angular velocity increases uniformly to 3·8 rads^{-1} in a time of 3·5 s. Show that

    a the angular acceleration is 0·34 rads^{-2}

    b the tangential acceleration is 1·4 ms^{-2}

    c the centripetal acceleration of the object is

        i 28 ms^{-2} at t = 0

        ii 59 ms^{-2} at t = 3·5.

CENTRIPETAL FORCE

Objects moving in a circular path will **accelerate towards the centre of the circle**. Newton's second law tells us there must be an **unbalanced force** causing this acceleration. This force is called the **centripetal force** or **central force**.

$$F = ma = m\frac{v^2}{r} = m\omega^2 r$$

Mass rotating horizontally on the end of a string

The tension in the string provides the centripetal force on the mass.

$$T = m\frac{v^2}{r}$$

A centripetal force is required to maintain circular motion. If the string breaks, the centripetal force will be removed and the mass will then move in a straight line, at a tangent to the circle (Newton I).

EXERCISE

2   Calculate the centripetal force on a mass of 0·4 kg making 3 revolutions per second in a circle of radius 0·5 m.

Spin dryer

Wet clothes move in a circle due to an inward force provided by the drum on the clothes. Water over the holes in the drum is not in contact with the drum, does not experience a centripetal force and moves tangentially in a straight line. It is a common misconception that the water will fly out radially due to some 'outward force'. This is not correct.

EXERCISE

3   A damp sock of mass 40 g experiences a centripetal force of 78·1 N from the drum of a washing machine on spin cycle. The radius of the drum is 22 cm. Calculate the angular speed of the drum in rpm on the spin cycle.

Aircraft turning (banking)

An aircraft in level flight has its weight balanced by the lift force **L**. When banking at angle θ, the horizontal component of **L** provides the centripetal force.

$$L\sin\theta = m\frac{v^2}{r}$$

Notice that **mg** > **Lcos**θ (vertical forces), so the aircraft may lose height when banking unless the lift **L** is increased by increasing speed.

Conical pendulum

A conical pendulum consists of a mass **m** on the end of a string. The mass travels in a horizontal circle as shown. The centripetal force is provided by the horizontal component of the tension **T** in the string.

$$T\sin\theta = m\frac{v^2}{r} = m\omega^2 r$$

$$T\cos\theta = mg \qquad \text{divide first equation by the second}$$

$$\frac{T\sin\theta}{T\cos\theta} = \frac{m\frac{v^2}{r}}{mg} = \frac{v^2}{rg}$$

$$\tan\theta = \frac{v^2}{rg} = \frac{r\omega^2}{g} \qquad \text{Also, } \sin\theta = \frac{r}{L} \text{ where } L \text{ is the length of the string.}$$

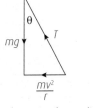

$\frac{mv^2}{r}$

Alternative forces diagram

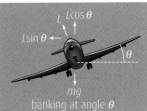

ONLINE

Visit www.brightredbooks.net to see how friction provides the centripetal force for a car travelling round a bend in the road.

ONLINE TEST

Head to www.brightredbooks.net and test your knowledge of centripetal acceleration and force.

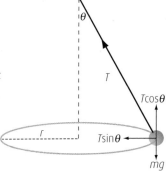

water
water droplet

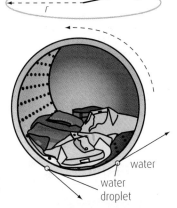

THINGS TO DO AND THINK ABOUT

There is no such thing as an outward force (centrifugal) acting on an orbiting mass. An object will fly out at a tangent to the circle if the central (centripetal) force is not great enough to keep it in that orbit. Many non-physicists incorrectly think this tangential motion is a radial motion and is due to an outward radial force which they call a centrifugal force.

ROTATIONAL DYNAMICS: MOMENT OF INERTIA

MOMENT OF INERTIA: AN OVERVIEW

The mass of a rotating object is an important consideration when calculating the **kinetic energy** of the object. The **distribution of this mass** about the axis of rotation is also an important factor.

An object rotates with **uniform angular velocity** ω about an axis through **O**.

Consider the object as a series of individual point masses m_1, m_2, m_3, ... at different distances from the axis of rotation, as shown in the diagram. The **kinetic energy** E_k of the rotating object is the **sum of the kinetic energies of each point mass**.

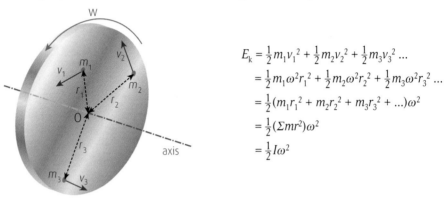

$$E_k = \tfrac{1}{2}m_1 v_1^2 + \tfrac{1}{2}m_2 v_2^2 + \tfrac{1}{2}m_3 v_3^2 \ldots$$
$$= \tfrac{1}{2}m_1\omega^2 r_1^2 + \tfrac{1}{2}m_2\omega^2 r_2^2 + \tfrac{1}{2}m_3\omega^2 r_3^2 \ldots$$
$$= \tfrac{1}{2}(m_1 r_1^2 + m_2 r_2^2 + m_3 r_3^2 + \ldots)\omega^2$$
$$= \tfrac{1}{2}(\Sigma m r^2)\omega^2$$
$$= \tfrac{1}{2}I\omega^2$$

DON'T FORGET ✚

These five expressions are given in the AH data booklet.

where I is the sum of all the mr^2 values for all the particles in the object.

I is called the **moment of inertia** of the object and is a measure of an object's **resistance to change in its rotation rate**. I has the unit **kgm^2**.

Extension

The moment of inertia I of a rotating object depends on the **mass** of the object and the **distribution of the mass** about the axis of rotation and can be calculated using integration.

MOMENTS OF INERTIA FOR DIFFERENT OBJECTS

Expressions for the moment of inertia for five different objects rotating about certain axes are given in the diagram.

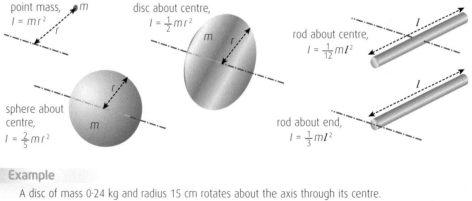

point mass, m
$I = mr^2$

disc about centre,
$I = \tfrac{1}{2}mr^2$

rod about centre,
$I = \tfrac{1}{12}ml^2$

sphere about centre,
$I = \tfrac{2}{5}mr^2$

rod about end,
$I = \tfrac{1}{3}ml^2$

Example

A disc of mass 0·24 kg and radius 15 cm rotates about the axis through its centre.
Calculate the moment of inertia of this disc.

Solution:

$I = \tfrac{1}{2}mr^2 = 0\cdot5 \times 0\cdot24 \times 0\cdot15^2 = 2\cdot7 \times 10^{-3}\,\text{kgm}^2$

contd

⚙ EXERCISE

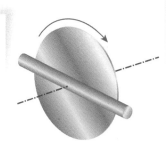

1   A rod of mass 0·21 kg and length 0·5 m is fixed to a disc of mass 0·12 kg and radius 0·18 m as shown. The rod–disc combination rotates about the axis at 15 rads⁻¹.

Show that the kinetic energy of the rod–disc combination is 0·71 J.

2   A sphere of mass 0·80 kg and radius 0·25 m rotates about an axis through its centre.

Calculate the moment of inertia of the sphere about this axis.

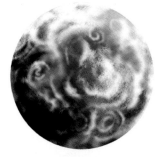

3   The Earth has a mass of $6·0 \times 10^{24}$ kg and a radius of $6·4 \times 10^{6}$ m.

   a   Calculate the angular velocity of the Earth in rads⁻¹.

   b   Calculate the moment of inertia of the Earth, assuming the Earth has uniform density.

   c   Calculate the kinetic energy of the rotating Earth.

4   A compact disc (CD) has a mass of 14 g and radius of 58 mm.

   a   Show that the moment of inertia of the CD is approximately $2·35 \times 10^{-5}$ kgm² about the central vertical axis.

   b   Show that the kinetic energy of the CD is approximately $5·9 \times 10^{-3}$ J when the angular velocity of the CD is 22·4 rads⁻¹.

   c   In practice, the CD has a central hole for fitting into the CD player. Will the moment of inertia of the CD be greater or less than $2·35 \times 10^{-3}$ kgm²? Justify your answer.

Absolute uncertainty in calculation of *I*

The formula for *I* involves a radius (or length) squared. The worked example shows how the uncertainty in the value of *I* is found.

Example

The mass and radius of a disc are measured

mass of disc, $m = 25 \pm 1$ g

radius of disc, $r = 8·0 \pm 0·2$ cm

Calculate the moment of inertia of the disc about its centre axis and the absolute uncertainty in the calculated value.

Solution:

$I = \frac{1}{2}mr^2 = 0·5 \times (25 \times 10^{-3}) \times (8·0 \times 10^{-2})^2 = 8·0 \times 10^{-5}$ kgm².

% uncertainty in $m = \frac{1}{25} \times 100 = 4\%$

% uncertainty in $r = \frac{0·2}{8} \times 100 = 2·5\%$

% uncertainty in $r^2 = 2 \times 2·5 = 5\%$

% uncertainty in $\frac{1}{2}mr^2 = \sqrt{4^2 + 5^2} = 6·4\%$

absolute uncertainty in I of disc = $6·4\% \times 8·0 \times 10^{-5} = 5 \times 10^{-6} = 0·5 \times 10^{-5}$ kgm²

$I_{disc} = (8·0 \pm 0·5) \times 10^{-5}$ kgm²

A complete treatment of uncertainties for AH Physics is covered in pages 116 to 121.

💭 THINGS TO DO AND THINK ABOUT

Revisit Example 3 above. In reality, the Earth is not a uniform sphere. The core of the Earth is denser than the region nearer the Earth's crust. What effect would this have on the magnitude of the moment of inertia of the Earth calculated above?

How would the kinetic energy of the rotating Earth be different from the value calculated above?

➕ DON'T FORGET

You must calculate *I* before using $E_K = \frac{1}{2}I\omega^2$

➡ ONLINE

Head to www.brightredbooks.net and check out the video on moment of inertia.

✔ ONLINE TEST

Take the test on the moment of inertia at www.brightredbooks.net

ROTATIONAL DYNAMICS: TORQUE

TORQUE: AN OVERVIEW

axis

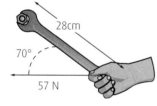

Torque T is the **turning effect** of a **force** on a **rotating object**. The disc can be made to turn about its **axis** by applying a **force F** at a **perpendicular distance r** from the axis of rotation. The turning effect on the disc will be less if distance r is reduced.

torque = applied force × perpendicular distance between direction of force and axis of rotation

$T = F \times r$ The **unit of torque** is **Nm**

axis

65N

0.20m

Torque is sometimes called the **moment of a force**.

Example

A force of 65 N is applied tangentially to a disc of radius 0·20 m. The disc can rotate about an axis through its centre. Calculate the applied torque.

Solution:

$T = F \times r = 65 \times 0·2 = 13 \, \text{Nm}$.

28cm

70°

57 N

⚙ EXERCISE

1 A torque of 26 Nm is applied to a nut using a spanner as shown. Calculate the force exerted by the operator.

2 A 57 N force is applied to a spanner at an angle of 70° as shown. The spanner is held 28 cm from the nut. Calculate the torque applied to the nut.

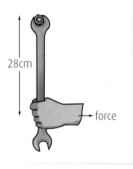

28cm

force

TORQUE AND ANGULAR ACCELERATION

An **unbalanced torque** on an object will produce an **angular acceleration** about an axis of rotation. The **angular acceleration** produced by the **unbalanced torque** depends on the **moment of inertia** of the object.

DON'T FORGET ➕

$T = I\alpha$ is the rotational motion equivalent of $F = ma$

The relationship linking **torque T** and **angular acceleration α** is

$T = I\alpha$ where I is the moment of inertia of the object about the axis.

This is the **rotational analogue** to Newton's second law $F = ma$

If a rotating object is subjected to friction, then a **frictional torque** will **oppose** the **angular acceleration**.

DON'T FORGET ➕

The symbol T in $T = I\alpha$ is the **unbalanced torque**.

unbalanced torque = applied torque − frictional torque

You are already familiar with finding the unbalanced force on an object. Unbalanced torque is found in a similar way.

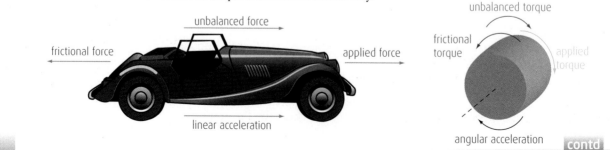

unbalanced force

frictional force

applied force

linear acceleration

unbalanced torque

frictional torque

applied torque

angular acceleration

contd

Example

A force of 15 N is applied to a string wrapped round the circumference of a disc of radius 0·30 m. The disc has a mass of 0·40 kg and accelerates at 35 rads⁻² about the axis.

Calculate

a the torque applied to the disc (by the string)

b the unbalanced torque on the disc

c the frictional torque acting on the disc.

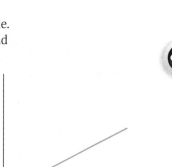

$a = 35\,\text{rads}^{-2}$

axis

15 N

Solution:

a Applied torque $= F \times r = 15 \times 0\cdot3 = 4\cdot5\,\text{Nm}$

b Calculate I first.
$I = \frac{1}{2}mr^2 = 0\cdot5 \times 0\cdot4 \times 0\cdot3^2 = 1\cdot8 \times 10^{-2}\,\text{kgm}^2$
Unbalanced torque $T = I\alpha = 1\cdot8 \times 10^{-2} \times 35 = 0\cdot63\,\text{Nm}$

c Unbalanced torque = applied torque − frictional torque
Frictional torque = applied torque − unbalanced torque $= 4\cdot5 − 0\cdot63 = 3\cdot87\,\text{Nm} = 3\cdot9\,\text{Nm}$.

Torque–angular acceleration graph

Circular motion can be investigated experimentally using a dedicated rotating low friction turntable. The pulley also has low-friction bearings.

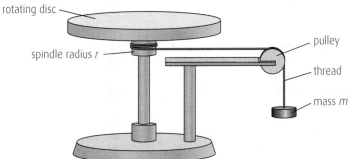

rotating disc

spindle radius r

pulley

thread

mass m

The torque is supplied by the thread wrapped round the spindle. Numerically, it is given by the product of the thread tension and the spindle radius

$T = \text{tension} \times r$

Increasing the mass hanging on the thread will increase the tension, so the torque increases.

A rotary motion sensor and computer software can measure the disc's angular velocity and acceleration.

Plotting torque T against angular acceleration α will give a straight line which should pass through the origin if friction is negligible.

The moment of inertia of the rotating disc and spindle can be worked out from the gradient of the straight line.

T

0

α

$I = gradient$

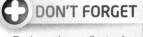

DON'T FORGET

$T = I\alpha$ so the gradient = I when T is plotted on the y-axis.

ONLINE

Learn more about the apparatus shown here at www.brightredbooks.net

THINGS TO DO AND THINK ABOUT

In the graph above, it is assumed that no frictional torque acts on either the turntable or pulley. If there had been a measurable frictional torque this would have caused a systematic uncertainty if the applied torque T was plotted as the unbalanced torque.

How would the graph above be affected by this systematic uncertainty? Can the frictional torque be estimated from the new graph?

ONLINE TEST

Test your knowledge of torque at www.brightredbooks.net

ROTATIONAL DYNAMICS: ANGULAR MOMENTUM

ANGULAR MOMENTUM: AN OVERVIEW

The **angular momentum** of a rotating object is defined as $L = I\omega$

The **unit of angular momentum** is **kgm²rads⁻¹** or **kgm²s⁻¹**.

This is the **rotational analogue** to **linear momentum** $p = mv$ where I replaces m and ω replaces v.

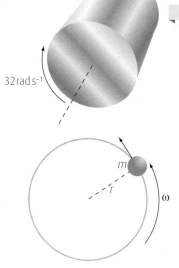

32 rad s⁻¹

Example

A uniform disc of mass 0·55 kg and radius 0·20 m rotates at 32 rads⁻¹.
Calculate the angular momentum of the cylinder.

Solution:

$$I_{disc} = \tfrac{1}{2}mr^2 = 0.5 \times 0.55 \times 0.2^2$$
$$= 1.1 \times 10^{-2}\,kgm^2$$
$$L = I\omega = 1.1 \times 10^{-2} \times 32 = 0.35\,kgm^2s^{-1}$$

Single particle moving in a circle

A single particle of mass m moving in a circle of radius r with angular velocity ω has an angular momentum as follows.

$$L = I\omega = mr^2\omega \text{ substituting } v = \omega r$$
$$L = mr^2\omega = mr^2\frac{v}{r} = mrv$$

The relationship in the relationships sheet written as $L = mrv = mr^2\omega$ refers to a **single particle only**. Solid objects are made of many single particles and the sum of each particle's angular momentum $\Sigma mr^2\omega$ gives the angular momentum of the continuous object.

CONSERVATION OF ANGULAR MOMENTUM

The **angular momentum** of a rotating object is **conserved** provided there are **no external torques** acting on the object.

0·13m

Example

A turntable of moment of inertia 6·2 × 10⁻³ kgm² rotates freely with no friction at 5·2 rads⁻¹. A small mass of 0·22 kg is dropped on to the turntable at a distance of 0·13 m from the axis of rotation. The mass stays at the same position after falling on to the turntable. Calculate the new angular velocity of the turntable and mass.

Solution:

$$I_{mass} = mr^2 = 0.22 \times 0.13^2 = 3.7 \times 10^{-3}\,kgm^2$$
Total angular momentum before = total angular momentum after

$$I_{turntable}\,\omega_1 = I_{turntable}\,\omega_2 + I_{mass}\,\omega_2$$
$$6.2 \times 10^{-3} \times 5.2 = (6.2 \times 10^{-3} + 3.7 \times 10^{-3})\,\omega_2$$
$$\omega_2 = 3.3\ rads^{-1}$$

DON'T FORGET

Remember to include $I\omega$ for the small mass.

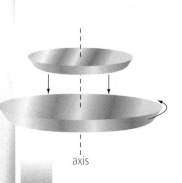

axis

⚙ EXERCISE

1 A disc of moment of inertia 7·5 × 10⁻³ kgm² rotates freely with no friction at 16 rads⁻¹. A second smaller disc of moment of inertia 4·2 × 10⁻³ kgm² drops on to the larger disc, with the centres of both discs on the axis of rotation.

   a Calculate the angular velocity of both discs after the impact, assuming both discs stick together and the smaller disc had no initial angular velocity.

   b Show that rotational kinetic energy is not conserved.

   c Calculate the amount of rotational kinetic energy converted into heat energy.

2 A horizontal disc rotates freely with negligible friction at 80 rpm. A small piece of putty of mass 2 × 10⁻² kg falls vertically on to the disc and sticks to it at a distance of 0·18 m from the axis of rotation. The disc plus putty now rotates at 70 rpm. Calculate the moment of inertia of the disc.

CONSERVATION OF ANGULAR MOMENTUM IN ACTION

Divers, ice skaters, acrobats and ballet dancers often make use of the conservation of angular momentum to increase or decrease their angular speed in a spin. A diver leaves the diving board with some initial angular velocity and with arms and legs outstretched. The axis of rotation is near the middle of her body. By pulling in her arms and legs, the diver's moment of inertia decreases as more of her mass is closer to the axis of rotation. Since I has decreased, then ω must increase to keep the angular momentum constant.

$L_{\text{initial}} = L_{\text{later}}$ (angular momentum conserved)

$I_{\text{big}}\, \omega_{\text{small}} = I_{\text{smaller}}\, \omega_{\text{bigger}}$ (with arms, legs pulled in)

The diver now stretches out her arms and legs as she approaches the water. This increases I and reduces ω.

Note that, although the force of gravity acts on the diver, this force does not exert a torque on the diver. The requirement of no external torques is met.

⚙ EXERCISE

3   An ice skater rotates at 3 rads^{-1} with her arms outstretched. Each arm has a mass of 6 kg, and her total arm span is 1·8 m. By treating both arms as a single rod, show that the moment of inertia of her outstretched arms is 3·2 kgm^2. The moment of inertia of the rest of her body is 0·6 kgm^2.

The skater now wraps her arms around her body. The arms can now be treated like a solid disc of radius 0·30 m.

   a   Show that the new angular velocity of the skater is 10 rad s^{-1}.

   b   Show that the rotational kinetic energy of the skater increases by 40 J when she pulls in her arms.

   c   Explain where the 40 J of kinetic energy has come from.

4   A bullet of mass 10 g travelling at 80 ms^{-1} strikes the edge of a stationary disc tangentially and lodges in it. The disc has a mass of 0·75 kg and a radius of 0·30 m and is free to rotate about an axis through its centre. Calculate

   a   the angular momentum of the bullet about the axis before impact

   b   the angular velocity of the disc and bullet after impact.

 ONLINE

Look up the link at www.brightredbooks.net for another description of angular momentum.

💭 THINGS TO DO AND THINK ABOUT

Models of global warming predict that large sections of the polar ice caps will melt.

Explain what effect this will have on the rotation of the Earth, however slight.

Your explanation should be based on physics principles and relationships (formulae).

✓ ONLINE TEST

Test your knowledge of angular momentum at www.brightredbooks.net

ROTATIONAL DYNAMICS: ROTATIONAL KINETIC ENERGY

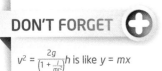

DETERMINE I OF CYLINDER ROLLING DOWN A SLOPE

A cylinder of mass m and radius r rolls down a slope from rest without slipping.

The top of the slope is at a height h above the bottom of the slope.

All the gravitational potential energy of the cylinder at the top of the slope will be changed to kinetic energy as the cylinder reaches the bottom of the slope.

The kinetic energy of the cylinder as it rolls down the slope consists of linear kinetic energy $\left(\frac{1}{2}mv^2\right)$ and rotational kinetic energy $\left(\frac{1}{2}I\omega^2\right)$

$$E_P \text{ at top of slope} = E_K \text{ at bottom of the slope}$$

$$mgh = \tfrac{1}{2}mv^2 + \tfrac{1}{2}I\omega^2$$

$$= \tfrac{1}{2}mv^2 + \tfrac{1}{2}I\frac{v^2}{r^2} \quad \text{since } \omega = \frac{v}{r}$$

$$mgh = \tfrac{1}{2}\left(m + \frac{I}{r^2}\right)v^2$$

$$gh = \tfrac{1}{2}\left(1 + \frac{I}{mr^2}\right)v^2 \quad \text{dividing through by } m$$

$$v^2 = \frac{2gh}{\left(1 + \frac{I}{mr^2}\right)} \quad \text{after rearranging}$$

A graph of v^2 against h should be a straight line through the origin with a gradient of $\frac{2g}{\left(1 + \frac{I}{mr^2}\right)}$.

The moment of inertia I of the cylinder rolling down the slope can be found by rearranging the expression for the gradient of the graph of v^2 against h.

$$\text{gradient} = \frac{2g}{\left(1 + \frac{I}{mr^2}\right)}$$

$$\left(1 + \frac{I}{mr^2}\right) = \frac{2g}{gradient}$$

$$\frac{I}{mr^2} = \frac{2g}{gradient} - 1$$

$$I = mr^2\left(\frac{2g}{gradient} - 1\right)$$

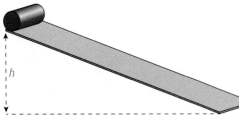

Substitute the values of cylinder mass m, radius r, the gradient of the graph and the value of g into this expression to determine the moment of inertia of the cylinder.

EXPERIMENTAL DETAILS

The height h of the top of the slope and the corresponding speed of the cylinder at the bottom of the slope can be measured. Repeat for various heights.

The speed of the cylinder at the bottom of the slope can be found using

- a motion sensor with related software and display,

- a light gate (care is needed to position the light beam to cross the diameter of the cylinder).

- average speed down slope $\bar{v} = \frac{distance \ down \ slope}{time \ taken \ down \ slope}$

speed at bottom of slope $v = 2 \times \bar{v}$

$$v = 2 \times \frac{distance \ down \ slope}{time \ taken \ down \ slope}.$$

UNIFORM SOLID CYLINDER

The moment of inertia of a uniform solid cylinder of mass m and radius r is given by the expression $\frac{1}{2}mr^2$. The moment of inertia found experimentally as described previously should agree with the moment of inertia calculated using $\frac{1}{2}mr^2$.

Experimental uncertainties can result in the theoretical and experimental results being different. If this experiment forms part of an investigation, then the report should discuss why the experimental and theoretical values for the moment of inertia of the cylinder are not the same. Here are some points to consider.

- How accurately was the height h measured and between which two points?
- Was the uncertainty in the gradient significant?
- Was the graph a straight line and did it go through the origin?

NON-UNIFORM CYLINDER

A plastic container filled with sand is rolled down a slope and the moment of inertia determined.

This cylindrical container approximates a uniform cylinder. The ends of the container confine more mass closer to the axis of rotation. This will reduce the value of I compared with a uniform cylinder of the same mass and radius as there is more mass at smaller values of r.

Example

A uniform cylinder of mass m and radius r rolls down a slope of height h. Show that the speed of the cylinder is given by the expression $v = \sqrt{\frac{4}{3}gh}$

Solution:

E_p at top of slope $= E_K$ at bottom of the slope

$mgh = \frac{1}{2}mv^2 + \frac{1}{2}I\omega^2$

$\qquad = \frac{1}{2}mv^2 + \frac{1}{2}\left(\frac{1}{2}mr^2\right)\frac{v^2}{r^2}$ $\qquad$ since $I = \frac{1}{2}mr^2$ and $\omega = \frac{v}{r}$

$mgh = \frac{3}{4}mv^2$ $\qquad$ divide both sides by m

$v^2 = \frac{4}{3}gh$

$v = \sqrt{\frac{4}{3}gh}$

 ONLINE

Explore this topic further by following the link at www.brightredbooks.net

⚙ EXERCISE

Show that the speed of a uniform sphere of mass m and radius r rolling down a slope of height h will have a speed at the bottom of the slope given by $\sqrt{\frac{10}{7}gh}$.

➕ DON'T FORGET

In a 'show' question all the physics must be shown.

💭 THINGS TO DO AND THINK ABOUT

A sealed cylindrical container full of water rolls down a slope. The same container full of ice now rolls down the slope. Which one will have the greater speed at the bottom? You may be able to test your hypothesis experimentally. Can you use your knowledge of physics to explain the result?

✓ ONLINE TEST

Head to www.brightredbooks.net and test yourself on this topic.

GRAVITATION: SCHIEHALLION EXPERIMENT

In 1774, the Astronomer Royal, Nevil Maskelyne, led an expedition to the Perthshire mountain, Schiehallion, to measure the density of the Earth and also to confirm Newton's law of gravitation between non-celestial masses on Earth.

Schiehallion has a symmetrical shape and is quite far away from other mountains, which made it an ideal choice for the experiment.

THEORY

A stationary pendulum held near the mountain would be deflected by a tiny amount, θ, towards the mountain due to the gravitational attraction between the mass of the mountain, M_s, and the mass of the pendulum bob, m.

The gravitational force of attraction, F, between m and M_s is:

$$F = \frac{GmM_s}{d^2}$$

The weight, W, of the pendulum bob is the gravitational force of attraction between the mass, m, and the mass of the Earth, M_E

$$W = \frac{GmM_E}{R_E^2}$$

The ratio of F to W gives

$$\frac{F}{W} = \frac{\dfrac{GmM_s}{d^2}}{\dfrac{GmM_E}{R_E^2}} = \frac{M_s R_E^2}{M_E d^2}$$

Substituting M_s using the density of Schiehallion $\rho_s = \dfrac{M_s}{V_s}$

and also substituting M_E using the density of the Earth $\rho_E = \dfrac{M_E}{V_E}$ gives

$$\frac{F}{W} = \frac{\rho_s V_s R_E^2}{\rho_E V_E d^2}$$

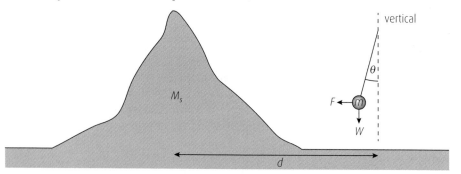

The tension T in the pendulum string can be split into components in the usual manner

$$F = T\sin\theta$$
$$W = T\cos\theta$$
$$\therefore \frac{F}{W} = \tan\theta$$

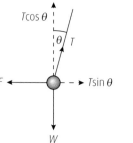

DON'T FORGET

$\dfrac{\sin\theta}{\cos\theta} = \tan\theta$

Substituting gives $\tan\theta = \dfrac{\rho_s V_s R_E^2}{\rho_E V_E d^2}$

Finally rearranging $\rho_E = \dfrac{\rho_s V_s R_E^2}{V_E d^2 \times \tan\theta}$ where the six quantities on the right hand side can all be measured.

contd

Extension – Taking the measurements

θ was the most difficult measurement as it was so small. Maskelyne used accurate astronomical instruments, clocks and his knowledge of star positions to measure his current latitude and then the vertical direction at this latitude. Next, he measured the direction of the stationary pendulum string. The difference between vertical and the pendulum string direction was θ. This was carried out at two locations north and south of Schiehallion to reduce uncertainties. His value of θ was 11·6 arcseconds, or $3\cdot2 \times 10^{-3}$ degrees. (An angle of $1°$ can be subdivided up into 60 arcminutes or 3600 arcseconds.)

The volume of Schiehallion was measured by the mathematician and surveyor Charles Hutton. He took thousands of bearings and developed the system of contour lines of equal height to work out the volume of individual sections before adding them up.

Charles Hutton also estimated the distance d from the pendulum bob to the centre of the mountain using all his survey points.

The density of Schiehallion was found using rock samples by dividing mass by volume. The average value was around $2500\,\text{kg}\,\text{m}^{-3}$ (in today's metric units).

The radius and volume of the Earth were known in 1774 as the Earth's circumference had been measured quite accurately by different methods over preceding centuries.

FACT

Venus subtends an angle between 10 and 50 arcseconds when seen from Earth.

ONLINE

Follow the link at www.brightredbooks.net for an excellent historical background to the experiment and also the difficulties faced by Maskelyne and his team.

AVERAGE DENSITY OF THE EARTH

Maskelyne's experiment gave a value for the density of the Earth of $4500\,\text{kg}\,\text{m}^{-3}$ which was almost twice the density of the mountain. This led to the hypothesis that the core of the Earth must be denser than the crust and must be metallic as metals are more dense than stone and rock.

Plaque at Schiehallion car park

Today, the accepted value of the density of the Earth is $5514\,\text{kg}\,\text{m}^{-3}$ so Maskelyne's results were within 20% of the accepted value.

Once the density of the Earth was found, then the mass of the Earth could also be found, using a calculated value for the volume ($m = \rho \times V$). This was the real prize as up to this point the masses of the Sun, planets and their moons could only be stated as a relative ratio with the Earth's mass. Now they could be calculated individually. Maskelyne's experiment is often referred to as 'weighing the Earth'.

This experiment also confirmed experimentally Newton's law of gravitation from 100 years earlier. Newton, at the time, had expressed the view that experimental verification of gravitational attraction would not be possible with non-celestial masses as the gravitational forces involved would be too small to measure.

denser core

THINGS TO DO AND THINK ABOUT

The mean value of the Earth's radius is 6371 km and the Earth's mean density is $5514\,\text{kg}\,\text{m}^{-3}$. Use these values to find the mean mass of the Earth. Why should your answer for the Earth's mass have four significant figures when using this data?

ONLINE TEST

Test yourself on this topic online at www.brightredbooks.net

GRAVITATION: CAVENDISH/BOYS EXPERIMENT

Almost 25 years after the Schiehallion experiment, in 1798, Henry Cavendish investigated another method of measuring the gravitational force between two masses.

Just under 100 years later, in 1895, Charles Boys further developed the Cavendish experiment and measured the value of *G* to within 0·2% of today's accepted value.

As a tribute to both scientists, the experiment carried out nowadays using a modern version of the apparatus is often called the Cavendish/Boys experiment.

THE CAVENDISH APPARATUS

Cavendish used a **torsion balance** consisting of a wooden rod suspended from a copper wire which could twist. The wooden rod had a small lead sphere at each end of the rod, and these small spheres were placed close to a large lead sphere. The wooden rod was 1·8 m long. The small spheres had mass 0·73 kg, and the large spheres had mass 158 kg.

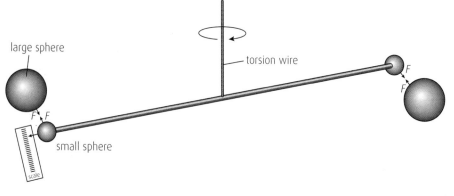

When the large lead spheres were brought close to the small spheres the gravitational force of attraction on the small sphere arrangement caused the suspended assembly to twist clockwise (as seen from above in the diagram). The tiny distance of arc moved by a pointer on one end of the wooden rod was measured against an accurate scale. The angle of twist could now be calculated using the relationship $angle\ of\ twist = \frac{arc\ length}{half\ rod\ length}$.

The whole assembly was housed inside a large shed with extensive draught-proofing. Cavendish observed the movement of the pointer against the scale from outside the shed using a powerful telescope. He also correctly reasoned that his own mass should be far away from the lead spheres to reduce gravitational forces due to his own mass.

ONLINE

Check out images of the original Cavendish apparatus at www.brightredbooks.net

THE BOYS APPARATUS

Boys' apparatus was similar to Cavendish's but the bar with the small spheres was suspended by a quartz fibre. The quartz fibre enables a greater angle of twist compared with a copper wire.

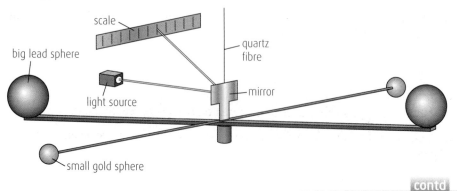

contd

The angle of twist was measured by recording the change in position of a light beam reflected from a mirror on to a scale.

Once the angle of twist was measured, the suspended assembly was allowed to oscillate and the period of oscillation T recorded.

Boys' apparatus had a period of oscillation of 2 to 3 minutes compared with the Cavendish version which had a period of 15 to 20 minutes.

Boys' apparatus also used a much smaller rod length with less massive spheres compared with the system used by Cavendish. This seems counterintuitive, but the improvement in the degree of twist of the quartz fibre compensates for the reduced gravitational force between the spheres.

Theory of Boys' method

The gravitational forces F on the smaller spheres at each end of the rod caused a torque on the suspended torsion balance.

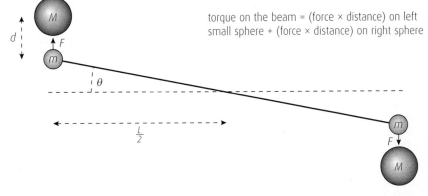

torque on the beam = (force × distance) on left small sphere + (force × distance) on right sphere

DON'T FORGET

There are two torques combining to produce an overall torque on the rod.

Torque $= \frac{GmM}{d^2} \times \frac{L}{2} + \frac{GmM}{d^2} \times \frac{L}{2}$ assumes F is perpendicular to the rod as θ is small

$\quad\quad\quad = \frac{GmML}{d^2}$

The torque is proportional to the torsion angle θ (or angle of twist)

$\frac{GmML}{d^2} \propto \theta$

$\frac{GmML}{d^2} = k\,\theta$ where k is the torsional constant of the wire

The suspended bar and small masses can oscillate freely with period T.

An expression for T can be found using knowledge of simple harmonic motion (see pages 70 to 75).

$T = 2\pi \sqrt{\frac{I}{k}}$ where I is the moment of inertia of the suspended bar and two small spheres.

Squaring both sides gives

$\quad k = \frac{4\pi^2 I}{T^2}$ and substituting $k = \frac{GmML}{d^2\theta}$ from above gives

$\frac{GmML}{d^2\theta} = \frac{4\pi^2 I}{T^2}$

$\therefore\ G = \frac{4\pi^2 I \theta d^2}{T^2 mML}$

All the terms on the right-hand side can be measured to find a value for G.

Boys found G to have a value of $6 \cdot 658 \times 10^{-11}\,\text{Nm}^2\text{kg}^{-2}$. This is within $0 \cdot 2\%$ of the current accepted value of $6 \cdot 67 \times 10^{-11}\,\text{Nm}^2\text{kg}^{-2}$.

THINGS TO DO AND THINK ABOUT

Show that the units on both sides of the equation $G = \frac{4\pi^2 I \theta d^2}{T^2 mML}$ are the same.

The unit of G is $\text{Nm}^2\text{kg}^{-2}$. Since $F = ma$, the Newton, N, has an alternative unit which is kgms^{-2}.

ONLINE TEST

Head to www. brightredbooks.net and test yourself on this topic.

GRAVITATION: GRAVITATIONAL FIELDS

NEWTON'S LAW OF GRAVITATION

The relationship $F = \frac{Gm_1m_2}{r^2}$ is called *Newton's law of Gravitation* and is studied in Higher Physics. The following revision exercise uses this relationship in an astronomical setting.

⚙ EXERCISE

1 The diagram shows the Earth and Jupiter orbiting the Sun.

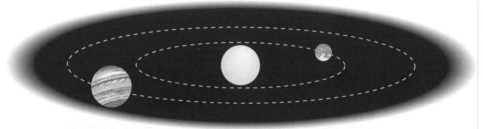

Astronomical data

Mass of Earth	$6 \cdot 0 \times 10^{24}$ kg
Radius of Earth	$6 \cdot 4 \times 10^6$ m
Mean radius of Earth orbit	$1 \cdot 5 \times 10^{11}$ m.
Mass of Moon	$7 \cdot 3 \times 10^{22}$ kg
Radius of Moon	$1 \cdot 7 \times 10^6$ m
Mean radius of Moon orbit	$3 \cdot 84 \times 10^8$ m
Mass of Jupiter	$1 \cdot 9 \times 10^{27}$ kg
Radius of Jupiter	$7 \cdot 0 \times 10^7$ m
Mean radius of Jupiter orbit	$7 \cdot 8 \times 10^{11}$ m

The gravitational force between the Earth and Jupiter varies as both planets orbit the Sun.

a Show that the maximum gravitational force between the Earth and Jupiter is $1 \cdot 9 \times 10^{18}$ N.

b Show that the minimum gravitational force between the Earth and Jupiter is $8 \cdot 8 \times 10^{17}$ N.

DON'T FORGET ➕

There are two radius values given for each object in the data table. Make sure the correct one is chosen.

GRAVITATIONAL FIELD STRENGTH

A **gravitational field** is a region where **gravitational forces** exist. **Gravitational field strength g** at a point is defined as the **force per unit mass** at that point. Or, more simply, it is the **force** on a **1 kg mass** placed at that point. In Higher Physics, you will have found that **g** on the surface of the Earth is **9·8 Nkg⁻¹**. This is consistent with the **inverse square law of gravitation**.

Gravitational force per kg on Earth's surface $= \frac{Gm_1m_2}{r^2}$

$= \frac{6 \cdot 67 \times 10^{-11} \times 6 \times 10^{24} \times 1}{(6 \cdot 4 \times 10^6)^2}$

$= 9 \cdot 8$ N (per kilogram)

g only has the value 9·8 Nkg⁻¹ at points on or near the Earth's surface. The value of **g** will **decrease** at a point high **above the Earth's surface** as **r increases**. The graph shows how **g** changes as **r** increases where **r** is the distance from centre of the Earth.

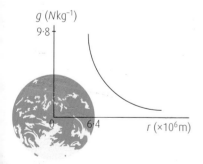

g (Nkg⁻¹)
9·8

0   6·4   r (×10⁶ m)

⚙ EXERCISE

2 Complete the table to show **g** decreasing with distance from centre of Earth.

r (× 10⁶ m)	6·4	8	10	15	20
g (Nkg⁻¹)	9·8				

3 Calculate **g** on the surface of the Moon.

GRAVITATIONAL FIELD LINES AROUND PLANETS.

The gravitational field around a mass is usually represented by lines which have the direction of the field shown by arrows on the lines. Each arrow shows the direction of the gravitational force acting on a 1 kg mass. A strong gravitational field has its field lines close together. The following diagram shows the gravitational field lines around the Earth.

Gravitational field lines around a planet.

The gravitational field around the Earth and Moon system is shown by the following diagram.

Earth

Moon

x $(3.84 \times 10^8 - x)$

Each gravitational field line shows the direction of the force experienced by a 1 kg mass. The pull of the Earth and Moon will result in two gravitational forces acting on the 1 kg mass. The gravitational field direction will show the resultant direction of the two gravitational forces.

The gravitational field strength at point X will be zero. The pull of the Earth and Moon on a unit mass at point X will be equal and opposite.

The exact position of point X can be calculated.

Let x be the distance between the centre of the Earth and point X.

The distance between point X and the centre of the Moon will be $(3.84 \times 10^8 - x)$

← gravitational force due to Earth = gravitational force due to Moon →

$$\frac{GM_{Earth} \times 1}{x^2} = \frac{GM_{Moon} \times 1}{(3.84 \times 10^8 - x)^2}$$ cancel G and take the square root of both sides to solve for x.

$x = 3.5 \times 10^8$ m from the centre of the Earth.

ONLINE

Learn more about this topic by following the link at www.brightredbooks.net

THINGS TO DO AND THINK ABOUT

Tides on Earth are caused by the force of gravity from the Moon and Sun acting on the oceans. The rotation of the Earth also plays an important role in the rise and fall of sea level.

The coasts of Scotland experience two high tides and two low tides every day (or just greater than 24 hours).

Some locations on Earth only experience one high tide every day. Carry out some research into why this happens and discuss in class, if time permits.

ONLINE TEST

Head to www.brightredbooks.net and test yourself on this topic.

GRAVITATION: SATELLITE MOTION

SATELLITE MOTION: AN OVERVIEW

Satellites are kept in orbit around a planet by the gravitational force, which provides the centripetal force on the satellite. Equating these two forces gives

$$\frac{Gm_p m_s}{r^2} = m_s \omega^2 r \quad m_p = \text{mass of planet}, m_s = \text{mass of satellite}$$

$$\frac{Gm_p m_s}{r^2} = m_s 4\frac{\pi^2}{T^2}r \quad \text{substituting } \omega = \frac{2\pi}{T} \text{ then cancelling } m_s \text{ gives}$$

$$T^2 = \frac{4\pi^2}{Gm_p}r^3 \quad \text{Kepler's third law (not in the relationships sheet)}$$

Kepler's third law states that the **period of a satellite squared** is **directly proportional** to the **cube of the radius of the orbit**. This relationship is very useful for calculations involving satellite orbits.

DON'T FORGET

The expression $\omega^2 = \frac{Gm_p}{r^3}$ can also be derived.

Example

A satellite orbits 300 km above the Earth's surface.

a Calculate the time taken for one orbit of the Earth.

b Calculate the speed of the satellite in this orbit.

Solution:

$r = r_{Earth} + \text{height above Earth}$

$= 6\cdot4 \times 10^6 + 300 \times 10^3$

$= 6\cdot7 \times 10^6 \, m$

$T^2 = \frac{4\pi^2}{Gm_p}r^3$

$= \frac{4 \times 3\cdot14^2 \times (6\cdot7 \times 10^6)^3}{6\cdot67 \times 10^{-11} \times 6 \times 10^{24}}$

$= 2\cdot96 \times 10^7$

$\Rightarrow T = 5\cdot4 \times 10^3 \, s = 90\cdot7 \text{ minutes.}$

b $\frac{m_{satellite}v^2}{r} = \frac{Gm_{Earth}m_{satellite}}{r^2}$ rearrange to get

$v = \sqrt{\frac{Gm_{Earth}}{r}} = \sqrt{\frac{6\cdot67 \times 10^{-11} \times 6 \times 10^{24}}{6\cdot7 \times 10^6}} = 7\cdot7 \times 10^3 \, ms^{-1}$

contd

EXERCISE

1 A weather satellite orbits the Earth every 87 minutes. Calculate the height of the satellite above the Earth's surface.

2 Calculate the height of a geostationary satellite above the Earth's equator.

3 The Moon is a satellite of the Earth. Show that the Moon takes just under one month to orbit the Earth.

4 Titan is the largest of Saturn's moons, with an orbit time of 15·9 days at a mean orbit radius of $1·22 \times 10^9$ m. Calculate the mass of Saturn.

5 The International Space Station (ISS) makes around 15·7 orbits of the Earth per day. Calculate the height of the ISS above the Earth's surface, assuming circular orbits.

6 Information about Jupiter's four Galilean moons is given in the table.

	Io	Europa	Ganymede	Calisto
Orbit time T (days)	1·77	3·55	7·16	16·7
Mean orbit radius R ($\times 10^8$ km)	4·22	6·71	10·7	18·8

Show that T^2 is directly proportional to R^3.

Galileo saw these moons changing position with his own basic telescope. He reasoned that the moons must be orbiting Jupiter. Sometimes, there were only three moons visible as one was either in front of or behind Jupiter. You too can share the experience and see these Galilean moons with a pair of ordinary binoculars.

7 Calculate the linear speed of the Earth as it orbits the Sun.

8 Show that the linear speed of an orbiting satellite is given by $v = \sqrt{\frac{GM}{R}}$ where M is the mass of the planet and R is the radius of the orbit.

THINGS TO DO AND THINK ABOUT

The distance from the Earth to the Sun is often quoted as 93 million miles (1 mile = 1609 metres). Calculate the number of days in a year using these figures.

Is the distance 93 million miles slightly bigger or smaller than the actual distance? Explain your answer.

DON'T FORGET

The orbit radius is the Earth's radius + the height of the satellite.

ONLINE

Check out the links at www.brightredbooks.net for information on Jupiter and to try more calculations on satellite orbits.

ONLINE TEST

Head to www.brightredbooks.net and take the test on satellite motion.

GRAVITATION: GRAVITATIONAL POTENTIAL

In earlier physics work, gravitational potential energy was calculated using the relationship $E_p = mgh$ where h is the vertical height above the Earth's surface. The potential energy at the Earth's surface was taken as zero, and the relationship is valid provided g remains constant. However, the force of gravity **does** change as we move away from the Earth, so a new expression for potential energy is required that can be applied at all points in space. To calculate **gravitational potential energy**, we need to introduce the new concept of **gravitational potential**, and use integration to calculate gravitational potential energy.

GRAVITATIONAL POTENTIAL ENERGY

DON'T FORGET

The minus sign must be included.

Gravitational potential at a point in space is defined as the **work done** by external forces to move **unit mass** from **infinity** to that point. **Infinity** is where the **force of gravity** is **zero**. The **gravitational potential** V at a **distance** r from a **mass** m is given by

$$V = -\frac{GM}{r}$$ the unit of V is Jkg^{-1}

GRAVITATIONAL POTENTIAL ENERGY

The **gravitational potential energy** of **mass** m_1 at a distance r from another mass m is given by $E_p = -\frac{Gmm_1}{r}$ the unit of E_p is J

DON'T FORGET

An alternative formula is $E_p = Vm$.

The minus sign appears during the integration and can be explained as follows: moving m_1 away from m requires an **increasing** amount of **work done**, overcoming the **attractive force** between the two masses. The **gravitational potential energy** must **increase to zero** at **infinity**, so it must be **negative** at **all points between m and infinity**.

Example

A satellite of mass 250 kg orbits the Earth at a height of 200 km. Calculate

a the gravitational potential at this height

b the gravitational potential energy of the satellite.

Solution:

a $R_{orbit} = R_{Earth} + height = 6\cdot4 \times 10^6 + 200 \times 10^3 = 6\cdot6 \times 10^6$ m

$$V = -\frac{GM}{r}$$

$$= -\frac{6\cdot67 \times 10^{-11} \times 6 \times 10^{24}}{6\cdot6 \times 10^6}$$

$$= -6\cdot06 \times 10^7 Jkg^{-1}$$

b $E_p = -\frac{Gmm_1}{r}$ or $E_p = Vm$

$$= -6\cdot06 \times 10^7 \times 250$$

$$= -1\cdot5 \times 10^{10} J$$

DON'T FORGET

Minus signs must be carried forward at all stages.

⚙ EXERCISE

1   Calculate the gravitational potential at a height of 150 km above the Moon's surface.

2   A satellite of mass 750 kg orbiting the Earth has a gravitational potential energy of $-4\cdot4 \times 10^{10}$ J. Calculate the height of the satellite above the Earth's surface.

ESCAPE VELOCITY

If a rocket is launched from Earth with sufficient kinetic energy to leave the Earth's gravitational field, it will not return. However, if the kinetic energy is not sufficient, then the rocket will fall back to Earth or stay in orbit round the Earth.

The **escape velocity** from a planet is defined as the **minimum velocity** required to just **escape from the planet's gravitational field** and **reach infinity** with **zero velocity**.

The derivation of escape velocity of a mass m from a planet of radius r is as follows.

Total energy on planet's surface = total energy at infinity = 0

$$E_k + E_p = 0$$

$$\frac{1}{2}mv^2 + \left(-\frac{GMm}{r}\right) = 0 \quad \text{where } M \text{ is the mass of the planet}$$

$$v^2 = \frac{2GM}{r} \Rightarrow v = \sqrt{\frac{2GM}{r}}$$

DON'T FORGET

It would be incorrect to begin with $E_k = E_p$.

Example

Calculate the theoretical escape velocity of a rocket fired from the surface of the Earth.

Solution:

$$v = \sqrt{\frac{2GM}{r}} = \sqrt{\frac{2 \times 6 \cdot 67 \times 10^{-11} \times 6 \times 10^{24}}{6 \cdot 4 \times 10^6}} = 1 \cdot 1 \times 10^4 \, \text{ms}^{-1}.$$

The effects of air resistance have not been included, and the rocket would need to reach this speed quite soon after launch. In practice, spacecraft deliberately escaping the Earth's gravitational field will be placed in orbit first (above the Earth's atmosphere) then given an escape velocity from the orbit (see Exercise 5).

VIDEO LINK

Check out the video at www.brightredbooks.net to explore this topic further.

 EXERCISE

3   An object of mass m escapes from a planet of mass M and radius r. Show that the minimum kinetic energy required is given by $E_k = \frac{GMm}{r}$

4   Calculate the escape velocity from the Moon.

5   Calculate the escape velocity from an orbit 300 km above the Earth's surface.

6   Calculate the escape velocity from a spherical asteroid of radius 150 km and density 2500 kgm⁻³.

ONLINE

What happens when speeds are greater or less than the escape velocity is illustrated in the examples at www.brightredbooks.net

SATELLITES CHANGING ORBIT

A satellite orbiting the Earth at a fixed radius has both kinetic and potential energy. The sum of these energies gives the total energy of the satellite in this orbit. The total energy of the satellite will be different in a higher orbit. The amount of energy required to change to a higher orbit is the difference between the total energy in each orbit.

 THINGS TO DO AND THINK ABOUT

Now have a go at this exercise on satellites changing orbit.

A satellite of mass $4 \cdot 5 \times 10^3$ kg orbits the Earth. The satellite increases its orbit radius from $6 \cdot 65 \times 10^6$ m to $6 \cdot 66 \times 10^6$ m.

a   Show that the satellite speed changes from $7 \cdot 758 \times 10^3$ ms⁻¹ to $7 \cdot 752 \times 10^3$ ms⁻¹.
b   Show that the kinetic energy of the satellite changes from $1 \cdot 354 \times 10^{11}$ J to $1 \cdot 352 \times 10^{11}$ J.
c   Show that the potential energy of the satellite changes from $-2 \cdot 708 \times 10^{11}$ J to $-2 \cdot 704 \times 10^{11}$ J.
d   Show that the total energy of the satellite changes from $-1 \cdot 354 \times 10^{11}$ J to $-1 \cdot 352 \times 10^{11}$ J.
e   Show that the (minimum) energy required to change orbit is $2 \cdot 0 \times 10^8$ J

ONLINE TEST

Head to www.brightredbooks.net and take the test on this topic.

GENERAL RELATIVITY: SPECIAL RELATIVITY

REVISION

Special relativity was studied in Higher Physics and deals with what happens when objects move at higher speeds than we would normally meet in everyday life.

To a stationary observer, a moving clock will run slower than a stationary clock. This is called **time dilation**. The moving clock shows time t' while the stationary clock shows time t. The relationship

$$t' = t \frac{1}{\sqrt{1 - \frac{v^2}{c^2}}}$$ must be used, where the symbols have their usual meanings.

To a stationary observer, the length l of a fast moving object will appear smaller. This is called **length contraction** and the relationship

$$l' = l \sqrt{1 - \frac{v^2}{c^2}}$$ must be used.

FRAMES OF REFERENCE

Two people moving at different speeds relative to each other are said to be in different frames of reference.

Consider a passenger in a bus passing a stationary observer. The bus has a steady speed of 20 ms^{-1}. Both observer and passenger have watches.

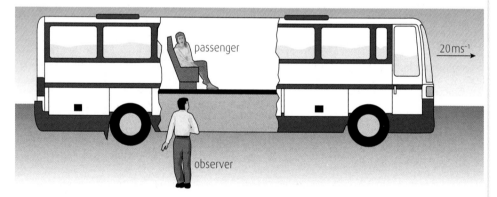

The observer sees the passenger moving at 20 ms^{-1} to the right.

The passenger sees the observer moving to the left at 20 ms^{-1}.

There are two frames of reference in the above diagram. One is centred on the observer and the other is centred on the passenger.

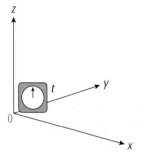

observer's frame of reference

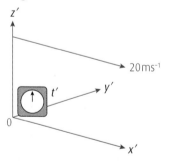

passenger's frame of reference

contd

There are four spacetime coordinates in each frame.

The spacetime coordinates for the observer are (x, y, z, t) and are (x', y', z', t') for the passenger.

If the passenger's steady speed was very high, then the observer would notice the passenger's watch running slowly and the dimensions of the passenger's watch would appear smaller as well.

Extension – Special Relativity timelines

In 1865 the Scottish physicist James Clerk Maxwell showed that electromagnetic radiation had a speed that was equal to a combination of two universal constants (see page 114). In other words, the speed of light also had a constant value. Physicists at the time concluded that this must be for a stationary frame of reference which they called the **ether**. As a consequence the speed of light would be greater or less in moving frames of reference.

In 1887, Michelson and Morley tried to measure the effect of the ether on the time a beam of light took to travel between a set of mirrors. They measured the same value of time regardless of the mirror orientation or which way the Earth was moving in its orbit around the Sun. This was not expected by the physics community and theoretical work began investigating why the speed of light appeared to be independent of the reference frame.

From 1890 to 1905, various scientists, including Lorentz, Fitzgerald, Poincairé and Larmor had realised that distance would have to be contracted (and time stretched) for Maxwell's equations to still be correct when going from the stationary (hypothetical) ether to a moving frame of reference.

By 1905 the Dutch physicist Hendrik Lorentz had developed a four-dimensional spacetime transformation which successfully showed distance and time changing when moving from his imagined stationary ether to a moving frame of reference. The Lorentz factor $\gamma = \frac{1}{\sqrt{1 - \frac{v^2}{c^2}}}$ was used in his theory.

In 1905 Einstein produced his paper which abandoned the ether idea altogether and derived the Lorentz transformation assuming the speed of light was constant in any inertial frame of reference.

INERTIAL FRAMES OF REFERENCE

Frames of reference which move at constant speeds relative to each other are said to be **inertial frames of reference**.

Einstein's theory of Special Relativity deals with inertial frames of reference only and states:

- the speed of light in each frame of reference is c regardless of how fast a frame was moving relative to the other frame

- the laws of physics apply equally to all inertial frames of reference

- at low speeds Special Relativity reduces to classical Newtonian physics

- mass and energy are equivalent ($E = mc^2$).

THINGS TO DO AND THINK ABOUT

The Lorentz factor $\frac{1}{\sqrt{1 - \frac{v^2}{c^2}}}$ is normally given the symbol γ and $\frac{v}{c}$ is given the symbol β.
This makes relativistic equations less cluttered and easier to write down and read.

The length contraction relationship $l = l' \sqrt{1 - \frac{v^2}{c^2}}$ can be written as $l = \gamma l'$

And γ can have the alternative form $\gamma = \frac{1}{\sqrt{1 - \beta^2}}$

DON'T FORGET

A watch running slowly means a time of 1 second has increased.

VIDEO LINK

Visit www.brightredbooks.net to view a video showing how frames of reference like the ones above can be used to develop the formulae for special relativity.

ONLINE

Visit www.brightredbooks.net and follow the link to 'evaluation of symbols' for a Lorentz factor calculator.

ONLINE TEST

Head to www.brightredbooks.net to test yourself on special relativity.

GENERAL RELATIVITY: EQUIVALENCE PRINCIPLE

After publishing his theory of Special Relativity, Einstein turned his attention to what happens when frames of reference are accelerating. These are called **non-inertial frames of reference**.

Einstein used a thought experiment to form the opinion that the force of gravity might be an illusion.

EINSTEIN'S THOUGHT EXPERIMENT (1907)

Einstein considered two scenarios.

Firstly a person in a windowless room on the Earth's surface drops a ball from rest.

Secondly, the same person is in a similar room inside a space craft in deep space accelerating upwards at $9.8\,\mathrm{ms^{-2}}$. The same ball is dropped.

acceleration = $9.8\,\mathrm{ms^{-2}}$

Both situations would result in the person experiencing exactly the same sensation of force. The ball dropped in each room would behave in exactly the same way. There would be no way of knowing whether the windowless room was stationary on Earth or accelerating in a region of zero gravity.

This realisation led Einstein to propose that acceleration and gravity are indistinguishable or equivalent.

EINSTEIN'S EQUIVALENCE PRINCIPLE

Einstein's equivalence principle states: *no observer can determine by experiment whether they are in an accelerating frame of reference or in a gravitational field.*

The laws of physics cannot distinguish between acceleration and gravity.

GRAVITATIONAL MASS AND INERTIAL MASS

The **gravitational mass** m of an object in a gravitational field can be found experimentally with these steps:

- measure the weight W of the object
- measure the value of g in the gravitational field
- use $W = mg$ to calculate the value of the mass m. This is the **gravitational mass.**

In a region of zero gravitational field the **inertial mass** m of the object can be found with these steps:

- apply an unbalanced force F to the mass
- measure the resulting acceleration a
- use $F = ma$ to calculate the mass m of the object. This is the **inertial mass.**

Einstein proposed that gravitational mass is the same as inertial mass.

BENDING OF LIGHT

Einstein reasoned that, since light has energy, it must also have an associated mass equivalent due to $E = mc^2$. As a consequence of the equivalence principle, a light beam will change direction when it passes through the gravitational field associated with a large mass.

The effect of the Earth's gravitational field on a beam of light is negligible since the Earth's gravitational field is very small. A large gravitational field will change the direction of light passing through it.

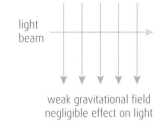

light beam

weak gravitational field
negligible effect on light

light beam

strong gravitational field
light beam changes direction

VIDEO LINK

Another early success for General Relativity is the explanation of why Mercury's orbit didn't exactly follow Newton's Laws of motion. Visit www.brightredbooks.net to view an excellent short video on this.

SOLAR ECLIPSE 1919

Experimental verification of light being bent by a gravitational field came in 1919 during a total eclipse of the Sun. The British astronomer Arthur Eddington photographed a star which should have been hidden behind the Sun during the eclipse.

The angle between the two star positions was very small and there was an associated 20% uncertainty but this was enough to confirm that the mass of the Sun had affected the direction of the light passing close by. This news made headlines across the world and Albert Einstein became widely known after this.

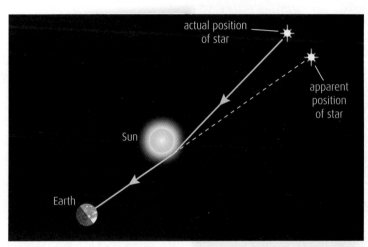

actual position of star

apparent position of star

Sun

Earth

GRAVITATIONAL LENSING

The Hubble telescope has photographed many distant galaxies with a ring of light around them. This is caused when a second galaxy lies directly behind the one being photographed.

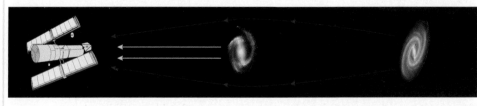

Light from the further galaxy is bent by the strong gravitational field of the nearer galaxy and is picked up by Hubble as a bright ring around the nearer galaxy's light. The ring of light is called an **Einstein Ring**.

Example of Einstein ring from Hubble

THINGS TO DO AND THINK ABOUT

Amazingly, Einstein was not awarded a Nobel Prize for his work on relativity or $E = mc^2$, but he did receive the Nobel Prize in 1922 for his earlier work on the photoelectric effect. Many scientists in the early twentieth century were sceptical of the as yet unverified theories of Special and General Relativity. Carry out an internet search on this and, if time, discuss your findings in class.

ONLINE TEST

Test your knowledge of stellar physics at www.brightredbooks.net

GENERAL RELATIVITY: CURVED SPACETIME

DON'T FORGET ➕

These changes in distance and time are separate from the special relativity changes due to speed.

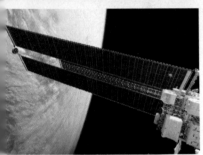

CURVED SPACETIME: AN OVERVIEW

Classical Newtonian physics treats three-dimensional space and time as separate entities. However, the theory of General Relativity is at odds with this approach.

Between 1907 and 1915 Einstein was trying to incorporate the concept of gravity into his Theory of General Relativity. He proposed that three-dimensional space and time are interconnected in a four-dimensional entity called **spacetime**.

Consider the following two non-relativistic motions
- a spacecraft in deep space moving at a steady speed
- the International Space Station (ISS) orbiting the Earth at a steady speed.

An astronaut present in either spacecraft will experience weightlessness. The spacecraft in deep space will travel in a straight line at a steady speed well away from any large mass.

The ISS travels in a curved path at a steady speed near a large mass (the Earth).

Einstein reasoned that four-dimensional spacetime would be affected by the presence of a large mass and concluded that spacetime would be curved near a large mass. The General Theory of Relativity states that both lengths and times will change near large masses and these changes in the spacetime coordinates can be interpreted as non-linear or curving.

Einstein described gravity as the bending or curving of spacetime geometry.

The General Theory of Relativity can show that a satellite will travel in a straight line in curved four-dimensional spacetime.

This is difficult to visualise but the following analogies are useful when thinking about curved spacetime.

RUBBER SHEET ANALOGY

Imagine a large mass placed on a horizontal rubber sheet. The rubber sheet has square grid lines on its surface representing a coordinate system (in two dimensions).

The sag in the middle is caused by the mass placed on the rubber sheet and the square grid lines are now curved near the mass. The grid lines remain unchanged in regions well away from the mass. The sag in the region around a large mass is sometimes called a **gravity well**.

This analogy is one way of trying to visualise how the presence of a mass can bend spacetime coordinates even although the analogy itself is not in four dimensions.

DON'T FORGET ➕

There is no time dimension in this analogy.

TWO-DIMENSIONAL MAP ANALOGY

A map of the Earth's surface is normally drawn on a flat (2D) page. This is not completely accurate, because the Earth is a sphere with three dimensions, so the map is a representation of a 3D region.

Consider a plane journey from Glasgow to New York.

Glasgow

New York

contd

The shortest distance between Glasgow and New York is drawn as a curved line on this 2D map.

The actual surface of the Earth is curved and not a flat 2D surface. This gives rise to the shortest journey between Glasgow and New York appearing as a curved line on the map with the aircraft continually changing direction.

SATELLITE MOTION

Let's go back to the ISS orbiting the Earth in a circular orbit. Einstein's theory of General Relativity suggests that the ISS is in fact moving in the straightest line allowed through the curved spacetime near the Earth.

We 'see' the ISS orbiting in a curved path because we are thinking in 3D not 4D and we are not taking into account that 4D spacetime is curved near the Earth.

In the flight path analogy above, we 'see' a curved path on the map because we are viewing a 3D journey in 2D. The aircraft is moving in the straightest line allowed between Glasgow and New York yet the 2D diagram shows a curved line.

The shortest journey between two points may be a straight line in 4D spacetime but it can appear to be a curved line if pictured in less than four dimensions. The shortest distance between two points in curved spacetime is called a **geodesic**.

THINGS TO DO AND THINK ABOUT

Einstein became something of an expert in the mathematics of curved space as he developed the General Theory of Relativity. The rules in curved space are different from those in the plane geometry we learn at school. To illustrate this, draw a triangle on a plastic ball. Measure all the angles with a protractor as best you can. Do they add up to 180°? Can you draw a triangle with three right angles on the surface of the ball?

ONLINE

Many good videos and animations have been made on curved spacetime. An internet search to find them will reinforce your understanding of this challenging topic.

ONLINE TEST

Head to www.brightredbooks.net and test yourself on curved spacetime.

GENERAL RELATIVITY: EFFECT OF ALTITUDE ON SPACETIME AND CLOCKS IN ACCELERATING SPACECRAFT

EFFECT OF ALTITUDE ON SPACETIME

General relativity predicts that spacetime becomes curved close to a large mass and less curved when further away from the mass. A consequence of this is that time will slow down or dilate close to a large mass compared to time measurements further away from the mass. Time will run slower under the influence of greater gravity.

Therefore on Earth, time will change with altitude above the Earth's surface.

The following diagram shows the (exaggerated) effect on two atomic clocks at different altitudes.

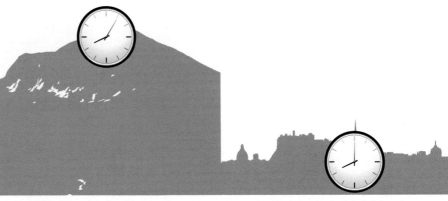

atomic clock on summit of Ben Nevis
(running faster)

atomic clock in Edinburgh
(running slower)

The city of Edinburgh is at a lower altitude than the summit of Ben Nevis so the gravitational field strength in Edinburgh is greater. Spacetime will be more curved in Edinburgh and an atomic clock in Edinburgh will run slower than a similar clock on Ben Nevis. The difference between the Edinburgh and Ben Nevis clocks will be so small that it would not normally be noticeable on ordinary clocks and watches. However, a time interval of one second in Edinburgh will be greater than a time interval of one second on Ben Nevis. A greater time of one second means that the clock will run slower.

General relativity also predicts that space (distance or length) will become smaller (contracted) near a large mass. This is quite separate from length contraction due to travelling at high speeds.

The following diagram summarises the effect of a large mass on space and time.

time dilating:
clocks running
slower

clocks running
faster

lengths
contracting

lengths
increasing

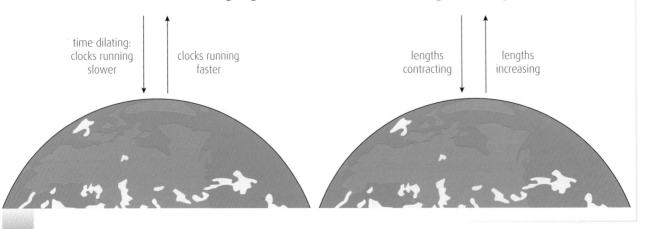

CLOCKS IN ACCELERATING SPACECRAFT

The equivalence principle states that a spacecraft accelerating at $9.8\,ms^{-2}$ in deep space will be indistinguishable from a similar stationary spacecraft on Earth. Consider the spacecraft at rest on Earth.

A clock at the top of the spacecraft will have a greater altitude than a clock at the bottom.

The clock at the top of the spacecraft will run faster than the clock at the bottom as previously discussed.

Consider now the same spacecraft in deep space accelerating forward at $9.8\,ms^{-2}$.

The equivalence principle states that all physics observations and experiments will be the same in both the stationary spacecraft on Earth and accelerating spacecraft.

Remember, an observer in the spacecraft (with no windows) will not know whether the spacecraft is stationary on Earth or accelerating in deep space at $9.8\,ms^{-2}$.

The clock at the front of the accelerating spacecraft will run faster than the clock at the rear.

As the acceleration of the spacecraft increases the time difference between the clocks will increase.

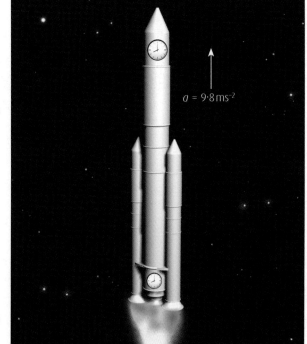

$a = 9.8\,ms^{-2}$

DON'T FORGET

A clock running slower means that time has been increased.

DON'T FORGET

That is provided the spacecraft accelerates at $9.8\,ms^{-2}$

ONLINE

Head to www.brightredbooks.net to watch a video on GPS and relativity corrections.

ONLINE TEST

Head to www.brightredbooks.net to test yourself on the effect of altitude on spacetime and clocks in accelerating spacecraft.

THINGS TO DO AND THINK ABOUT

Imagine, for a moment, a physics class in the future which has hand-held atomic clocks for student use. The clocks are accurate to 1 ns and can be zeroed and synchronised.

Use your knowledge of physics to describe an experiment you might be able to carry out with these clocks which would be impossible to do before the arrival of these clocks?

This could be described as an open-ended question as a variety of answers are possible. Open-ended questions appear in all physics exams today and are usually signposted by the expression 'use your knowledge of physics'.

GENERAL RELATIVITY: SPACETIME DIAGRAMS

SPACETIME DIAGRAMS: AN OVERVIEW

Einstein's Theories of Relativity deal with motion in four-dimensional spacetime. In 1908 Hermann Minkowski showed it is possible to illustrate many journeys through spacetime on a simple two dimensional diagram without the need for complicated mathematics.

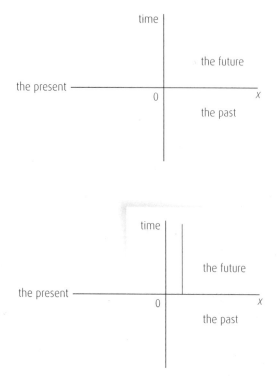

A spacetime diagram is really a graph. The horizontal axis represents the x-coordinate. The other two spatial coordinates are not shown.

Time is shown on the vertical axis. This is unlike most other graphs in physics which usually have time on the horizontal axis. Having time on the vertical axis means that everything **below the x-axis is the past** and everything **above the x-axis is the future**. Time $t = 0$ can be considered to be the **present**. Physicists often use the term *here and now* to describe the origin of the graph rather than 'the present'. The term 'here and now' recognises both position and time.

WORLDLINE

A line drawn on a spacetime diagram is called a **worldline**. Here is one example of a worldline.

This worldline drawn in red represents an object at rest. Its x-coordinate stays the same as time progresses into the future. The red line can be extended backwards into 'the past' if the object's position has remained unchanged for some time.

Worldline for an object at rest

LIGHT WORLDLINES

The units on the axes of a spacetime diagram often have time in years and x in light years (ly).

An observer at O sends a light ray in the $+x$-direction and simultaneously sends another light ray in the opposite direction.

Light travels a distance of 1 ly in a time of 1 year. Each ray of light therefore has a worldline at 45° to both the x-axis and t-axis as shown by the yellow lines on the spacetime diagram. Each light worldline can be extended back into the past.

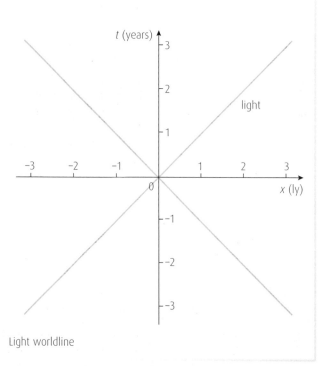

Light worldline

MOTION WORLDLINES

The following spacetime diagram shows the worldlines of moving objects.

The blue worldline represents an object moving at a steady speed. The x-coordinate increases evenly as time increases.

The orange worldline represents an object with a higher steady speed than the blue line. The object travels a greater distance in a fixed time than was the case with the blue line.

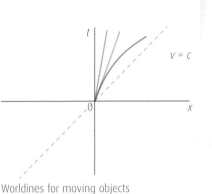

Worldines for moving objects

The green curved worldline represents an accelerating object.

As nothing can exceed the speed of light there can be no worldlines with a gradient less than 1 as this would mean travelling faster than the speed of light.

⚙ EXERCISE

A futuristic spacecraft sets off on a journey at a speed of $0\cdot1$c at time $t = 0$. On a spacetime diagram sketch the worldline of this journey. Some numerical values are required on the axes.

SPACETIME EVENT

A particular point in a spacetime diagram is called an **event**. Three events A, B and C are shown on the following spacetime diagram. An observer is at the origin O.

Event A has happened in the observer's past and is possibly recorded in history.

Event B will happen in the observer's future. This may be a supernova explosion and light from this will reach the observer (or successor on Earth) at some point in the future.

Event C lies outside the light worldlines. Light from event C would have to travel at faster than the speed of light to reach the observer. Event C will not be in the observer's past and cannot be recorded in history.

Extension – Relativistic effects

Spacetime diagrams come into their own when a second inertial frame of reference moves at a relativistic speed compared to a stationary observer. The resulting spacetime diagram will have two sets of axes (x, t) for the stationary observer and (x', t') for the fast-moving frame of reference (shown in red).

The observer at O is just about to send out two pulses of light, 1 second apart (blue lines). Both pulses will reach the fast-moving frame of reference (red) and will appear to the observer at O to have a separation of greater than 1 second. This is consistent with time appearing to slow down in the fast-moving frame when viewed by the stationary observer. A longer recorded time for 1 second is equivalent to clocks running slower.

The diagram also shows the speed of light being constant in both inertial frames as the same dotted line is used in both frames of reference. This is consistent with the speed of light being constant in all reference frames.

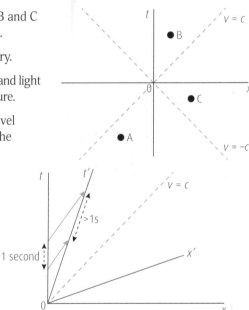

Spacetime diagram with two frames of reference

💭 THINGS TO DO AND THINK ABOUT

There are many examples on the internet of spacetime or Minkowski diagrams of relativistic motion. Check some of them out and see if you can follow the logic.

Einstein was initially dubious of Minkowski's graphical approach to relativistic motion but eventually recognised the merits of simple spacetime graphs as he developed his theory of General Relativity. A young Einstein had been one of Professor Minkowski's students at university.

✚ DON'T FORGET

In conventional speed-time graphs the less steep gradient line represents a lower steady speed.

✚ DON'T FORGET

A straight line at 45° to the x- and t-axes has a gradient of 1.

➡ ONLINE

Carry out an online search for physics videos to reinforce the basics of spacetime diagrams.

✚ DON'T FORGET

The blue lines are 1 second apart in the stationary frame of reference and are parallel to the v = c line.

✓ ONLINE TEST

Test yourself on spacetime diagrams at www.brightredbooks.net

GENERAL RELATIVITY: BLACK HOLES

BLACK HOLES: AN OVERVIEW

A black hole is formed when a massive star runs out of fuel for fusion and collapses in on itself. The process is similar to the creation of a neutron star described on page 56. If the mass of the star is big enough a black hole will result rather than a neutron star. The density of a black hole is huge as a large mass has been concentrated into a small volume.

A black hole can be described in terms of gravitation, and in terms of curved spacetime.

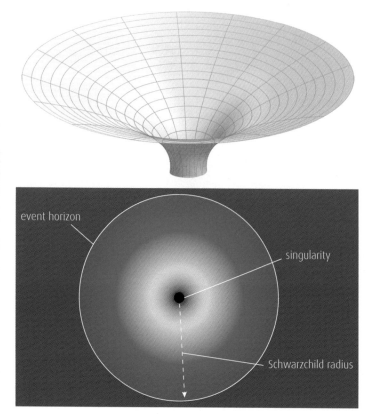

The gravitational description of a black hole states that the intense gravitational field at the surface of the black hole is so large that the escape velocity exceeds the speed of light. Nothing, including light, can escape from the surface of a black hole.

Einstein's theory of General Relativity suggests that what we perceive as gravity close to a large mass is in fact the curvature of four-dimensional spacetime. The greater the mass, the greater the curvature of 4D spacetime will be.

A black hole has infinite curvature of spacetime. The rubber sheet analogy suggests using a diagram like this in an attempt to visualise the curving of spacetime by a black hole.

The current theory of black holes predicts that the mass of the black hole will reduce in size to a point called a **singularity**.

The **event horizon** is the boundary between the inside of a black hole and the outside universe. The escape velocity from the event horizon of the black hole is equal to the speed of light. The event horizon forms the surface of a sphere, the radius of which is called the **Schwarzschild radius**.

SCHWARZSCHILD RADIUS

The Schwarzschild radius is the radius of a spherical mass where the escape velocity from the surface of the mass is equal to the speed of light. The relationship for the Schwarzschild radius r_{sch} is

$$r_{sch} = \frac{2MG}{c^2}$$

where M = the mass of the spherical object
G = the gravitational constant
c = the speed of light

Example

Calculate the Schwarzschild radius of the Sun.

Solution:

$$r_{sch} = \frac{2MG}{c^2} = \frac{2 \times (2 \cdot 0 \times 10^{30}) \times (6 \cdot 67 \times 10^{-11})}{(3 \cdot 0 \times 10^8)^2}$$

$$= 2960\,m$$

$$= 3 \cdot 0 \times 10^3\,m$$

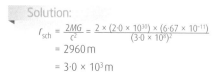

contd

Significant figures (s.f.)

The three values of numerical data used in this worked example have a mixture of 2 significant figures (for M and c) and 3 significant figures (for G). The final answer should really have only 2 significant figures. However, there would be no penalty for using 3 significant figures in the answer above. Writers of exam questions usually make sure any numerical data used all have the same number of s.f. where possible. This avoids confusion, and the final answer should have this same number of s.f. You are not penalised for having one or two extra s.f. in a final answer. You can also have one less s.f. than the final answer without penalty. Writing down the full string of digits on your calculator for your **final** answer will be penalised (unless all the data has 10 s.f., which is very unlikely).

You will also be penalised for incorrect rounding in a final answer, such as rounding 2960 m to 2.9×10^3 m in the above example.

SOME HISTORY OF BLACK HOLES

Karl Schwarzschild's work in 1916 on finding the radius of the event horizon was considered a curiosity at the time and the term 'black hole' had not been developed. The term **frozen star** was used at one point as it was predicted (correctly) that time would stand still at the event horizon.

The term 'black hole' is credited to physicist John Wheeler in 1967, although a journalist, Ann Ewing, had used it in the title of one of her science-related articles a couple of years earlier.

The discovery of the first neutron star in 1969 reignited interest in this branch of cosmology.

Stephen Hawking was one of many physicists to study the mathematics and physics of black holes even although none had been positively identified at that time. Hawking's work led him to propose that black holes emit radiation called **Hawking radiation**. This proved controversial at the time as nothing was supposed to escape from a black hole.

CYGNUS X-1

Cygnus X-1 is an X-ray source in the constellation Cygnus (6000 ly from Earth) and was the first candidate to be proposed as the source of a black hole. It is part of a binary system and has been studied intensively ever since its discovery in the 1960s. Its mass is estimated to be 15 times the mass of the Sun and it has a radius of 44 km. By 1990, Cygnus X-1 was confirmed as a black hole although all the evidence is indirect.

ONLINE

Photographs of Cygnus X-1 taken by NASA are quite unremarkable although a NASA artist's impression of what is happening there is quite fascinating. Search for 'NASA Cygnus X-1 image' and visit the NASA page for the explanation.

THINGS TO DO AND THINK ABOUT

A supermassive black hole is believed to be at the centre of the Milky Way. Its mass has been estimated to be 4·1 million Solar masses. Calculate its Schwarzschild radius. Now work out the density of this supermassive black hole assuming it is spherical and the mass is evenly spread throughout its volume.

Calculate the density of the Cygnus X-1 black hole.

What happens to the density of black holes as the mass increases?

ONLINE TEST

Test yourself on this topic at www.brightredbooks.net

STELLAR PHYSICS: LUMINOSITY AND APPARENT BRIGHTNESS

The stars we see from Earth have many properties such as age, size, brightness and temperature. Some properties are intrinsic to the star itself while other properties are those observed from here on Earth.

We will study the following properties of stars and how they are interconnected

- luminosity
- surface temperature

and also as seen from Earth

- apparent brightness
- distance from Earth to the star.

LUMINOSITY

The luminosity L of a star is the total energy radiated per second by the star. This can also be expressed as the total power radiated by the star.

The unit of luminosity is the watt W or Js^{-1}.

The Sun has a luminosity of 3.85×10^{26} W. It is sometimes useful to compare the luminosity of a star with that of the Sun. Astronomers use the subscript $\odot$ when referring to the Sun so a common abbreviation for the luminosity of the Sun is

$L_\odot = 3.85 \times 10^{26}$ W.

The luminosities of some other stars are listed in the table.

star	luminosity (W)	luminosity ($L_\odot$)
Vega	1.54×10^{28}	40
Betelgeuse	4.62×10^{31}	1.20×10^5
Proxima Centauri	6.5×10^{23}	<1%

APPARENT BRIGHTNESS

The night sky from Earth reveals stars of varying brightness.

The apparent brightness of a star observed on Earth is related to the luminosity of the star and the distance from Earth to the star.

The total energy per second (or total power) emitted by the star radiates out into space until some of this energy per second reaches Earth.

contd

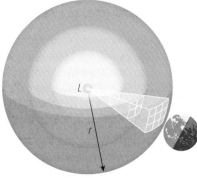

1 m² area on Earth's surface

The apparent brightness of a star is the amount of energy per second from the star landing normally on 1 square metre at the surface of the Earth.

The total power L is spread out over the surface area of ever-increasing spheres as the distance from the star increases.

each square represents an area of 1 square metre

The fraction of the star's total power, L, landing normally on an area of $1\,m^2$ on the surface of the outer sphere in the diagram is $\dfrac{L}{surface\ area\ of\ sphere}$ which is $\dfrac{L}{4\pi r^2}$.

The apparent brightness of a star has the symbol b and is defined as the amount of energy per second (or power) from the star landing normally on 1 square metre at the surface of the Earth.

Relationship $b = \dfrac{L}{4\pi r^2}$

A star of luminosity L and distance r from the Earth has an apparent brightness b given by

$b = \dfrac{L}{4\pi r^2}$ The unit of b is Wm^{-2}.

Example

Vega has a luminosity 40 times greater than the Sun and is 25 light years (ly) from Earth. Calculate the apparent brightness of Vega.

Solution:

$b = \dfrac{L}{4\pi r^2}$

$= \dfrac{40 \times (3.85 \times 10^{26})}{4 \times 3.14 \times (25 \times 365 \times 24 \times 60 \times 60 \times 3 \times 10^8)^2}$

$= 2.2 \times 10^{-8}\ Wm^{-2}$

DON'T FORGET

The surface area of a sphere of radius r is $4\pi r^2$.

ONLINE

Explore this topic further by following the link at www.brightredbooks.net

THINGS TO DO AND THINK ABOUT

The Greek astronomer Hipparchus (190–120 BC) classified stars according to their brightness using nothing more than the naked eye as his guide. The brightest stars were given magnitude +1 and the dimmest stars were +6. A modified version of this magnitude system is still used today by astronomers. Polaris (the pole star) has a magnitude of +1.97 and the planet Uranus has a magnitude of +5.6 (barely visible to the naked eye).

Really bright celestial objects would be outside the original range and have negative magnitudes, for example the full moon has a magnitude of –13. Discuss whether you think this seems counterintuitive.

ONLINE TEST

Want to revise your knowledge of this topic? Test yourself online at www.brightredbooks.net

STELLAR PHYSICS: LUMINOSITY AND STEFAN–BOLTZMANN LAW

SURFACE TEMPERATURE OF A STAR

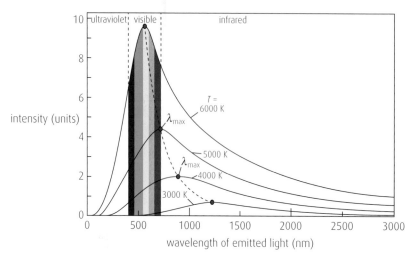

A star radiates energy and the intensity of the energy emitted depends on the surface temperature of the star and the wavelength of the light emitted. The graph shows the range of wavelengths of light emitted at various surface temperatures. The peak wavelength λ_{max} at each temperature is where the intensity of the emitted light is greatest.

The changing shape of the graphs can be explained. More energy is radiated as the star's surface temperature increases. More energy corresponds to a higher frequency of radiation ($E = hf$) and smaller wavelength ($\lambda = \frac{v}{f}$).

Wien's Law states that λ_{max} is inversely proportional to surface temperature T or

$$\lambda_{max} = \frac{2 \cdot 9 \times 10^{-3}}{T} \quad \text{where } \lambda_{max} \text{ has units in m and } T \text{ has units in K.}$$

By analysing light from a star, the maximum wavelength λ_{max} can be identified. Wien's Law can then be used to calculate the surface temperature T of the star.

Extension

Wein's law is not in the syllabus, but could arise under problem-solving.

STEFAN–BOLTZMANN LAW

All objects radiate energy. Experiments show that the total energy emitted per second, P, from the object depends on the surface temperature T (K) and surface area A of the object.

For perfect radiators $\qquad P \propto A$
$\qquad\qquad\qquad$ and $\qquad P \propto T^4$

Combining gives the relationship

$P = \sigma A T^4$ where σ is the Stefan–Boltzmann constant with a value of $5 \cdot 67 \times 10^{-8}\,\text{Wm}^{-2}\text{K}^{-4}$.

Josef Stefan used experimental data to show that the power emitted from a hot wire was proportional to the fourth power of its surface temperature. His postgraduate student Ludwig Boltzmann developed the theory of this relationship. The relationship $P = \sigma A T^4$ is called the **Stefan–Boltzmann Law** in recognition of the work done by both scientists.

Part of the experimental data used by Stefan is reproduced below.
A red-hot platinum wire at 525°C radiated 11·4 units of power.
The same white-hot platinum wire at 1200°C radiated 122 units of power.
Can you follow in Stefan's footsteps and show how this data pointed to a fourth power relationship?

DON'T FORGET

This data quoted here would have been part of a much larger set of data used by Stefan.

LUMINOSITY, SURFACE TEMPERATURE AND RADIUS OF A STAR

A star is a hot sphere radiating energy and the total power P radiated by the star is given by

$P = \sigma A T^4$

The surface area of a spherical star is $4\pi r^2$ where r is the radius of the star and T is the surface temperature of the star in Kelvin.

The total power radiated by a star is $P = 4\pi r^2 \sigma T^4$

The total power emitted by the star is also the luminosity L of the star.

This gives the relationship for the luminosity L of a star

$L = 4\pi r^2 \sigma T^4$

Example

The Sun has a luminosity $3\cdot83 \times 10^{26}$ W and a surface temperature of 5780 K.

Calculate the radius of the Sun.

Solution:

$$L = 4\pi r^2 \sigma T^4$$
$$r^2 = \frac{L}{4\pi\sigma T^4}$$
$$r = \sqrt{\frac{L}{4\pi\sigma T^4}}$$
$$= \sqrt{\frac{3\cdot83 \times 10^{26}}{4 \times 3\cdot14 \times 5\cdot67 \times 10^{-8} \times 5780^4}}$$
$$= 6\cdot94 \times 10^8 \text{ m}$$

The relationship $L = 4\pi r^2 \sigma T^4$ has three variables to different powers and this introduces the possibility of slightly more complex data handling calculations as shown in the following worked example.

Example

The radius of star A is three times that of star B.

The surface temperature of star A is twice that of star B.

How does the luminosity of star A compare with that of star B?

Solution:

$$\frac{L_A}{L_B} = \frac{4\pi \times (R_A)^2 \times \sigma \times (T_A)^4}{4\pi \times (R_B)^2 \times \sigma \times (T_B)^4} = \frac{4\pi \times \left(\frac{R_A}{R_B}\right)^2 \times \sigma \times \left(\frac{T_A}{T_B}\right)^4}{4\pi \times \sigma} = \left(\frac{3}{1}\right)^2 \times \left(\frac{2}{1}\right)^4 = 9 \times 16 = 144$$

Star A is 144 times more luminous than star B.

EXERCISE

1   The luminosity of star X is three times that of star Y.
    The surface temperature of star X is half that of star Y.
    Show that the radius of star X is approximately seven times that of star Y.

2   The luminosity of star P is twice that of star Q while the radius of star P is one third of star Q.
    Show that the surface temperature of star P is just over twice that of star Q.

THINGS TO DO AND THINK ABOUT

The surface temperature of sunspots is around 4000 K compared to the normal surface temperature of the Sun at 5700 K. Using the physics developed on this page, show that sunspots emit 27% of normal Sun luminosity.

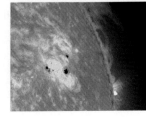

 ONLINE

Carry out an internet search for details of how Stefan calculated the surface temperature of the Sun from an experiment using a hot metal disc. His value of 5430°C (5700 K) is surprisingly accurate.

 **DON'T FORGET**

The final answer has 3 s.f. as all the data used in the calculation has 3 s.f.

 **DON'T FORGET**

This is a 'show' question, so trial-and-error substitutions until a number close to 7 is found will not be acceptable. The relevant formula and substitutions must be included in a 'show' question.

 **ONLINE TEST**

Test your knowledge of stellar physics at www.brightredbooks.net

STELLAR PHYSICS: STELLAR DISTANCES

THE LIGHT YEAR

We have seen in Higher Physics that the metre is not really a convenient unit of distance in astronomy due to the vast distances involved. The light year (the distance travelled by light in one year) is much more convenient as it turns huge distances into more manageable values. Proxima Centauri, the closest star to the sun, is 4·2 ly away ($3\cdot99 \times 10^{16}$ m).

THE ASTRONOMICAL UNIT

Another unit of distance used by astronomers is the Astronomical Unit (AU).

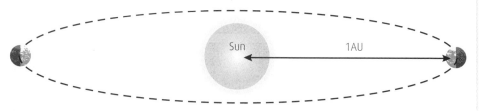

1 AU = average distance between the Earth and Sun ($1\cdot5 \times 10^{11}$ m)

The AU is most convenient when discussing distances in the Solar System. For example, Venus is 0·72 AU from the Sun while Jupiter is 5·2 AU from the Sun.

Extension

Distances using parallax

Parallax effects can be used to measure distances to many of the stars visible in the night sky.

A star viewed from one position will be seen in a unique direction from the observer. When the star is viewed from a different position it will now have a different direction. The angle between these two directions can be used to calculate the distance to the star.

A star is viewed in early January and its direction noted. Six months later in early July the star's new direction is noted. In these six months the Earth will have moved half a circumference as it orbits the Sun. The Earth's position in July will be will be a distance of 2 AU from its position in January.

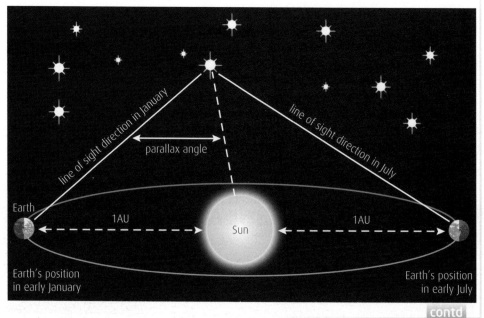

contd

The difference between the two angular directions is halved to give the parallax angle p.

Extension

The following theory on the parsec is not in the AH physics syllabus but could feature in data handling examination questions where appropriate information is supplied.

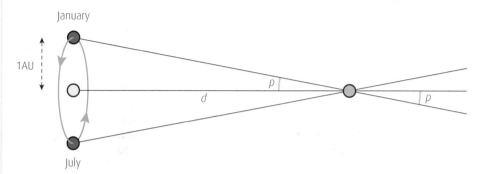

The angle p is very small and is measured in arcseconds.

An arcsecond is $\frac{1°}{3600}$ or $2·78 \times 10^{-4}$ degrees or $4·85 \times 10^{-6}$ radians.

$$\tan p = p = \frac{1(AU)}{d}$$

$$d = \frac{1}{p}$$

The unit of d is the **parsec** which is a combination of the words parallax and arcseconds.

A distance of 6·5 parsec means the parallax angle p is 6·5 arcseconds (over a six months period). Converting 6·5 parsec to metres is shown as an example of data handling.

$$6·5 \text{ parsec} = \frac{1}{p} = \frac{1AU}{6·5 \text{ parsec}} = \frac{1·5 \times 10^{11}(m)}{6·5 \times 4·85 \times 10^{-6}(rad)} = 4·8 \times 10^{15} m$$

Astronomical distance calculations are quick and easy using this parallax method and the parsec is the preferred unit for many astronomers and astrophysicists.

 EXERCISE

Show that 1 parsec is equivalent to

a $3·1 \times 10^{16}$ m.

b   3·3 ly

c $2·06 \times 10^5$ AU

THINGS TO DO AND THINK ABOUT

Popular science fiction normally prefers the light year as a measure of celestial distance. The Star Trek series, however, regularly used the term parsec correctly and in context. On the other hand, an early Star Wars film had Han Solo (a young Harrison Ford) claiming his spacecraft "had done the Kessel Run in less than 12 parsecs". Sounds very much like wrong physics although various attempts have been made to explain that the statement was in fact valid. The arguments are well worth an internet search. Check these arguments out and discuss – wrong physics or not?

DON'T FORGET

$\tan \theta = \theta$ for small angles measured in radians.

DON'T FORGET

All the steps must be included in 'show' type examination questions.

ONLINE

Learn more about this topic by following the link at www.brightredbooks.net

ONLINE TEST

Test your knowledge of this topic online at www.brightredbooks.net

STELLAR PHYSICS: HERTZSPRUNG-RUSSELL DIAGRAMS

HERTZSPRUNG-RUSSELL DIAGRAMS

The Hertzsprung–Russell (H–R) diagram is a scatter graph that plots the luminosity of a star against its surface temperature. When hundreds of known stars are entered on the H-R diagram the results are not a random scatter but fall into four distinct groups: main sequence, giants, supergiants and white dwarfs.

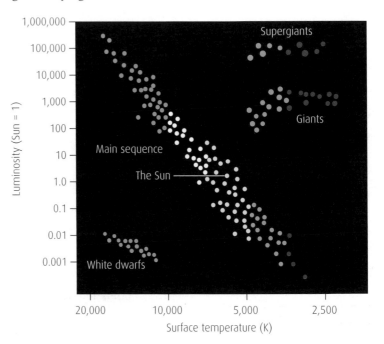

Note the temperature axis scale increases from right to left rather than the more conventional left to right and the scales on the axes are not linear.

Main sequence stars account for 90% of the stars in the Universe. These stars fuse hydrogen nuclei into helium nuclei. The Sun is a main sequence star with coordinates (5600, 1) on the above diagram.

Giant stars and supergiant stars have much bigger radii and luminosities than main sequence stars of the same surface temperature and are found in the top right of the H–R diagram. All the hydrogen in the core has been fused into helium and the helium in turn now fuses into carbon.

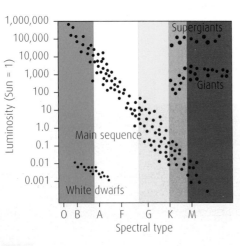

White dwarfs are the final evolutionary state of most stars once fusion has stopped. The very hot core of the former giant star has no source of energy and will gradually cool as it radiates heat energy.

More detail on the evolution of stars is given on pages 52–57.

The original H–R diagram was created by Ejnar Hertzsprung and Henry Russell working independently around 1911–12. Hertzsprung's horizontal axis used the colour of light emitted by each star. Russell used letters identifying the spectral type of the light given off by each star. The letters O, B, A, F, G, K and M were used which are meaningful and useful to astronomers, but surface temperature of the star is perhaps more user-friendly for our purposes.

The H–R diagram opposite shows how spectral type and the colour of light emitted can be used as alternatives to star surface temperature.

COMPARING STARS WITH THE SAME SURFACE TEMPERATURE

Three named stars are shown on the following H–R diagram.

Betelgeuse and Barnard's star have approximately the same surface temperature. It follows that they must have similar power outputs per square metre of surface area.

But Betelgeuse is much more luminous than Barnard's star so it must have a greater surface area to account for this greater overall power output. Betelgeuse is a much bigger star with more surface area than Barnard's star.

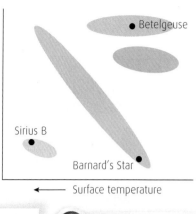

Betelgeuse

Sirius B

Barnard's Star

Luminosity

Surface temperature

COMPARING STARS WITH THE SAME LUMINOSITY

Barnard's star and Sirius B have similar (low) luminosities. But Sirius B is much hotter with a greater surface temperature. The power output per unit surface area of Sirius B is much greater than that of Barnard's star so Sirius B must have a much smaller surface area to account for this.

Sirius B has a much smaller radius and smaller surface area than Barnard's star.

Extension – Sirius A

Sirius A is the brightest star in the sky, as seen from Earth, and is a main sequence star with surface temperature of 9940 K and luminosity 25 times greater than the Sun ($25L_\odot$).

Sirius A is usually just called Sirius as its near neighbour, Sirius B, is so dim that it cannot be seen with the naked eye from Earth.

Sirius is called the 'dog star' as it is part of the Canis Major (great dog) constellation just below Orion (the hunter).

Sirius can be found by approximately projecting the line of Orion's belt to the lower left.

Sirius is relatively close to Earth at a distance of 8·6 ly which also contributes to it brightest star status.

In ancient Roman times Sirius could still be seen in the sky at sunrise between mid-July and mid-August. The expression 'dog days of summer' (meaning sultry and oppressive hot weather) comes from the time when the dog star could still be seen as the sun was rising.

Betelgeuse

Orion

Rigel

Sirius

Canis Major

This photo was taken from the Hubble telescope showing Sirius B as a faint dot to the lower left of Sirius A. The photograph of Sirius A was overexposed to enable Sirius B to be seen.

⚙ EXERCISE

1 In which region of the H–R diagram would you find
   a a hot star with a low luminosity (a hot dim star)
   b a cool star with high luminosity (a cool bright star)?

2 A star is hotter than the Sun and with a higher luminosity. What can be deduced about the radius of the star compared with the radius of the Sun?

ONLINE

For interactive questions on the H–R diagram, go to www.brightredbooks.net

☁ THINGS TO DO AND THINK ABOUT

Red dwarf stars form a subdivision of the main sequence at the bottom right of the H–R diagram. They are cooler, dimmer, smaller and less massive than the Sun. Proxima Centauri, the nearest star to the Sun, is a red dwarf. Red dwarf stars are predicted to live for trillions of years before turning into blue dwarfs. Blue dwarf stars are hypothetical at the present time and none have been observed. Can you think why none have been found yet?

ONLINE TEST

Head to www.brightredbooks.net and test yourself on this topic.

STELLAR PHYSICS: EVOLUTION OF THE SUN 1

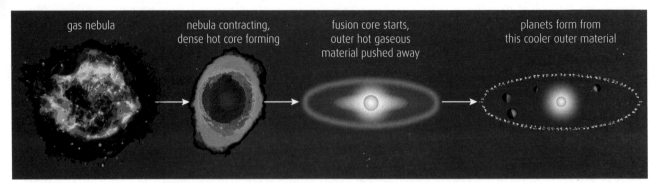

gas nebula

nebula contracting, dense hot core forming

fusion core starts, outer hot gaseous material pushed away

planets form from this cooler outer material

BIRTH OF THE SUN

The Sun started life as a nebular cloud of interstellar gas and dust consisting mainly of hydrogen. The force of gravity on the gas particles and dust caused them to move closer together forming a central core. A solar event like a nearby supernova explosion may have kickstarted this process. As the particles became closer together, the density of the core in the gas cloud gradually increased. This process would have taken many thousands of years.

The gas cloud would be spinning and the rate of spin would increase as more material moved closer to the core. This was due to conservation of angular momentum as the moment of inertia of the gas cloud would decrease as more mass moved to the core. The angular velocity would therefore increase to keep $I\omega$ constant.

The increasing number of collisions between the hydrogen nuclei in the core caused the temperature of the core to increase and it would become hot, visible and radiating energy. The core at this stage is called a **protostar**.

At some point, the temperature of the core was so great that fusion of hydrogen nuclei into helium started. The outer gas cloud was forced away when this happened leaving the core exposed as a new star – the Sun. The planets were formed from the particles in the outer gas cloud.

H–R DIAGRAM: PROTOSTAR SUN

The developing protostar will change position in an H–R diagram until it joins the main sequence as a stable new star. The red line on the H–R diagram shows how the developing protostar changes position.

The surface temperature of the developing protostar will increase as the core temperature rises.

The luminosity will decrease as the protostar is still contracting. When hydrogen starts to fuse into helium the luminosity increases slightly until the newly created star becomes stable.

luminosity

temperature

GRAVITATIONAL EQUILIBRIUM

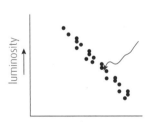

gravitational forces

outward thermal pressure forces

The hydrogen nuclei in the Sun gain energy from the heat generated by the fusion reaction and move faster. The Sun stops contracting as the inward gravitational forces are now balanced by the increased outward thermal pressure forces of the fast moving hydrogen nuclei. This balance is called **gravitational equilibrium**.

The Sun was born four-and-a-half billion years ago and is about halfway through its stable phase of gravitational equilibrium. The Sun will remain as a main sequence star on the H–R diagram during this time.

PROTON-PROTON CHAIN FUSION

Hydrogen nuclei in the Sun undergo nuclear fusion into helium nuclei releasing energy as some of the hydrogen mass is converted into energy.

The nuclear fusion reaction is not a simple single reaction but is a series of three separate fusion reactions called the **pp chain**.

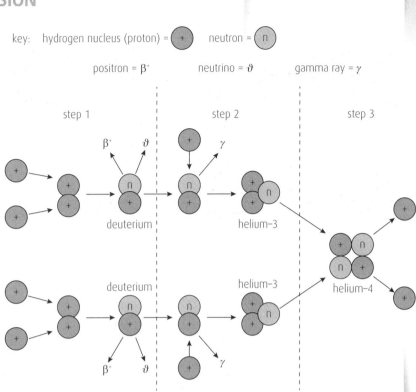

Step 1

Two hydrogen nuclei (protons) combine to form a deuterium nucleus which contains one proton and one neutron. A positron (β^+) and a neutrino (ϑ) are released at the same time.

This happens twice.

Step 2

The deuterium nucleus fuses with a proton to produce helium-3 or $_2^3 He$. A gamma ray is released at the same time.

This happens twice.

Step 3

Two helium-3 nuclei fuse to form helium-4 or $_2^4 He$. Two protons are released at the same time.

This three-stage reaction can be summarised.

$$4\,_1^1 H = \,_2^4 He + 2e^+ + 2\vartheta + 2\gamma$$

The mass of the products is less than the mass of the four initial protons and the missing mass is converted into energy by $E = mc^2$.

The energy released in the above reaction is 26·7 MeV.

In one second 600 million tonnes of hydrogen become 596 million tonnes of helium with four million tonnes of hydrogen converting directly into energy by $E = mc^2$.

⚙ EXERCISE

Show that four million tonnes of hydrogen converting to energy in the Sun each second is consistent with the numerical value of the Sun's luminosity.

💭 THINGS TO DO AND THINK ABOUT

Stage 1 of the pp chain has, surprisingly, a very low reaction rate but, once it takes place, stages 2 and 3 follow on relatively quickly. There are so many hydrogen nuclei in the Sun that the tiny proportion which fuses every second is enough to produce a huge amount of energy.

If the fusion reaction rate of stage 1 was greater, then more fusion reactions would take place in one second and the Sun would be much hotter. What effect do you think this would have on the lifetime of the Sun as a main sequence star?

➕ DON'T FORGET

Deuterium is an isotope of hydrogen with the symbol $_1^2 H$.

➕ DON'T FORGET

1 tonne = 1000 kg

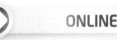

➡ ONLINE

Learn more about the evolution of the sun by following the link at www.brightredbooks.net

✔ ONLINE TEST

Test yourself on stellar physics at www.brightredbooks.net

STELLAR PHYSICS: EVOLUTION OF THE SUN 2

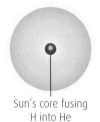

Sun's core fusing
H into He

helium core (no fusion)

pp reaction in hydrogen shell

THE NEXT 5 BILLION YEARS

The Sun at present is in a stable state fusing hydrogen into helium inside its core and releasing energy at a constant rate. This stable state will continue for another 5 billion years.

The Sun's position on the H–R diagram will stay unchanged as a main sequence star until most of the hydrogen has been converted into helium by the pp nuclear reaction.

RED GIANT BRANCH

Eventually most of the Sun's core will convert to helium and the pp reaction in the core will stop. However, the hydrogen in a thin shell surrounding the core will continue fusing to helium, releasing energy and at the same time adding more helium to the core.

The force of gravity will contract the core, which will heat up. This additional heat will cause the rate of fusion in the shell above the core to increase. The outward thermal pressure forces will be greater than the inward gravity forces and the volume of the Sun will increase.

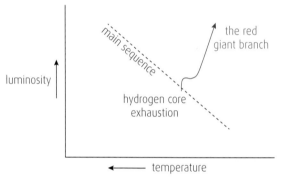

The Sun will leave the main sequence and the luminosity will increase. The surface temperature of the Sun will gradually decrease as its outer surface area increases and cools slightly.

The Sun will become a red giant with its outer edge at the orbit of Venus. The Sun will take a billion years to climb the red giant branch of the H–R diagram.

HELIUM CORE FUSION

As the Sun becomes a red giant, the mass of the inert helium core will continue to increase and gravitational forces continue contracting the core. The temperature of the core will rise until the helium in the core will begin fusing into carbon when the temperature reaches 100 million K. This point is known as the **helium flash**.

The helium fusion in the Sun will be a two-step chain called the **triple-alpha process** as three helium nuclei combine to make one carbon nucleus.

The reaction can be summarised as

$$^4_2He + {}^4_2He = {}^8_4Be + \gamma \quad \text{then} \quad {}^8_4Be + {}^4_2He = {}^{12}_6C + \gamma$$

The mass of the products is less than the mass of the 3 helium nuclei and this missing mass is converted into energy by $E = mc^2$.

There are now two sources of energy in the red giant Sun

* helium fusion in the core

* hydrogen fusion in the shell above the core.

contd

This extra energy will cause the temperature of the Sun to increase and the Sun's movement on the H–R diagram will change direction. The Sun's outer layers will contract and the luminosity decreases as the Sun moves back towards the main sequence region.

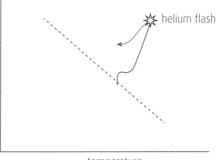

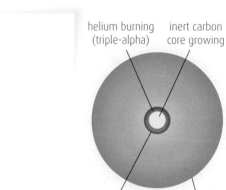

helium burning (triple-alpha)   inert carbon core forming

pp reaction in hydrogen shell

ASYMPTOTIC GIANT BRANCH PHASE

As the core of carbon builds up, triple-alpha fusion is limited to the shell above the core. Above this, there is still a shell of hydrogen fusing to helium.

The outward thermal pressure will increase and the Sun's radius will increase again and its surface will cool. The luminosity will increase due to the larger surface area.

helium burning (triple-alpha)   inert carbon core growing

pp reaction in hydrogen shell   outer edge growing

This second movement up the H–R diagram towards the red giant area is called the **asymptotic giant branch phase**.

END OF SUN'S FUSION

Once the fusion of helium and hydrogen ends, the Sun will have an extremely hot core and the outer envelope of the Sun will cool and slowly drift away as a cloud of cooling material called a **planetary nebula**. This eventually exposes the hot dense core which will radiate energy with no source of internal energy to replace it.

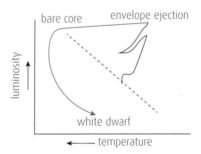

The outer nebula will drift away and the core will become exposed. The surface temperature of what is left of the Sun will increase until the core is completely exposed.

The hot core will now cool to become a white dwarf with a size approximately equal to the Earth.

ONLINE

Check out the clip 'The life and death of our Sun' at www.brightredbooks.net

ONLINE TEST

Head to www.brightredbooks.net and test your knowledge of stellar physics.

THINGS TO DO AND THINK ABOUT

The term planetary nebula is a bit of a misnomer as it has nothing to do with planets. The term was coined in the late eighteenth century by the astronomer William Herschel. He observed a cloud of gas around a star and thought this was the birth of a star with planets forming from the nebula. Herschel's name for this was adopted and has never been changed.

This photograph of a planetary nebula was taken by the Hubble telescope. Ultraviolet radiation from the hot core ionises the gas in the nebula resulting in the glowing appearance.

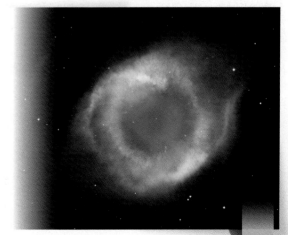

STELLAR EVOLUTION: NEUTRON STARS AND SUPERNOVA

NEUTRON STARS

Stars with a mass greater than $10\,M_\odot$ will have much greater internal gravitational forces resulting in higher core temperatures. These large stars initially behave like the Sun, fusing their hydrogen first then fusing helium (at a faster rate than the Sun). However, when the carbon core is formed this too will fuse, owing to the extremely high core temperatures.

The newly created nuclei will in turn fuse, resulting in the formation of even more new elements.

Eventually, there will be an inert core of iron with multiple shell fusion surrounding the core.

key: ● proton   ● neutron   · electron

Each iron nucleus in the core contains 26 protons and 30 neutrons.

There will be electrons in the hot core plasma as well.

Gravitational forces will cause the nuclei and electrons to come closer and closer together.

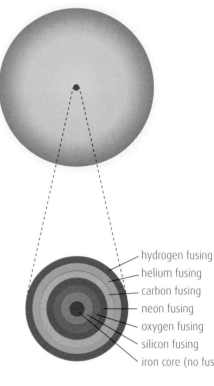

hydrogen fusing
helium fusing
carbon fusing
neon fusing
oxygen fusing
silicon fusing
iron core (no fusion)

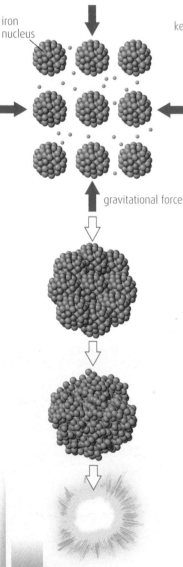

iron nucleus

gravitational force

The iron nuclei are forced closer together to within touching distance of each other.

Electrons will be present in the mix

The core temperature continues to increase.

The electrons combine with protons forming neutrons and releasing neutrinos

The star explodes creating a supernova where the material outside the core is blown away.

The diagram shows the composition of the various shells.

The core now consists of mainly iron which does not release energy due to nuclear fusion and so there is now no heat released by fusion in the core. The inward gravitational forces are no longer balanced by the outward thermal pressure forces and the core contracts.

The gravitational forces in the core become so great that the core collapses in a process that takes only seconds. The iron nuclei are forced closer and closer together and separate into protons and neutrons. Electrons present in the core combine with protons to produce neutrons, releasing a neutrino. The following diagrams show this process.

The extremely hot core of neutrons remains intact as a **neutron star**. It is very dense as there are no empty spaces between the neutrons.

Neutron stars have very large rotation rates due to the conservation of angular momentum. As more mass moves into the core, the moment of inertia decreases. The angular velocity increases to keep $I\omega$ constant.

contd

Some neutron stars emit radiation that can be detected on Earth and these are known as **pulsars**.

The nebula formed by the supernova will drift away and become the gas and dust required for the birth of new stars.

Spectral analyses of the light from supernovae shows the presence of elements heavier than iron and up to uranium. These heavier elements were produced during the supernova where the extra energy caused fusion of some of the elements already in the shells.

Gold is one of the elements created during a supernova when lighter nuclei combined to form gold nuclei. Gold is very unreactive and, once created, will survive in its metallic form. It is quite amazing to think that all the gold in a gold ring was formed during a supernova.

Crab Nebula

RECORDED SUPERNOVA

The most spectacular supernova in recorded history occurred in 1054 AD and is called SN 1054. This supernova would have been seen by anyone looking at the sky on a clear night.

SN 1054 was four times brighter than Venus and was visible in the night sky for almost two years. It could even be seen during daylight for the first 23 days. The Chinese astronomers who first observed it kept detailed records of this 'guest' star and more recently astronomers have identified the Crab Nebula as the remnants of SN 1054.

Photographs of the Crab Nebula over many years show that it is expanding. Astronomers have been able to use these photographs and their dates to help extrapolate backwards to when this nebula was created, to confirm the date of 1054 AD.

The most recent supernova visible to the naked eye was observed in 1987 (named SN 1987A). It was not as spectacular as SN 1054 and looked like an ordinary star. Neutrino detectors on Earth picked up an increased detection rate on the day SN 1987A was first observed.

ONLINE

Learn more about neutron stars by following the link at www.brightredbooks.net

 THINGS TO DO AND THINK ABOUT

Another spectacular supernova in our lifetime would be a truly memorable sight. When Betelgeuse eventually explodes as a supernova it is predicted to be as bright as the full Moon. Betelgeuse is over 600 ly from Earth. How many centuries will elapse between Betelgeuse's eventual supernova and it being observed here on Earth?

ONLINE TEST

Test yourself on this topic online at www.brightredbooks.net

INTRODUCTION TO QUANTUM THEORY

CHALLENGES TO CLASSICAL THEORY

Classical physics refers to any physics theory which survived successfully for a while but was ultimately replaced with a newer, more complete, theory. The newer theory could explain experimental anomalies which the classical theory could not.

Newtonian mechanics is an example of classical physics which is very successful at speeds well below the speed of light. At speeds above $0.1c$ however Newtonian mechanics is not accurate. Einstein's theory of special relativity has replaced classical physics when the speeds involved are large.

BLACK BODY RADIATION

Black body radiation is the name given to the radiation emitted by hot objects. A black body is an ideal perfect radiator. Classical physics treats black body radiation as a wave but the classical wave theory cannot explain some of the experimental evidence associated with this radiation.

The classical physics relationship governing the black body radiation emitted by a hot object is the Rayleigh–Jeans law. This states that the irradiance I of the radiation is inversely proportional to the fourth power of the wavelength λ of the radiation.

$$I \propto \frac{1}{\lambda^4}$$

This did not agree with experimental evidence especially at small wavelengths.

Experimental evidence shows that the irradiance peaks at one particular wavelength before falling to zero as the wavelength decreases.

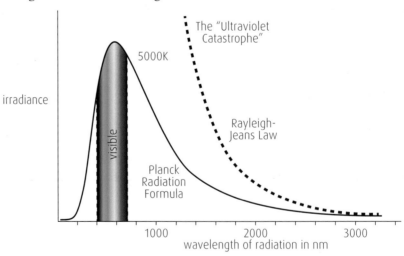

DON'T FORGET

Irradiance is the power received per unit area. $I = \frac{P}{A}$ The unit of I is Wm^{-2}.

The Rayleigh–Jeans theory predicted that the irradiance would increase continuously (as shown by the dotted line on the graph). As the wavelength λ decreases, $\frac{1}{\lambda^4}$ will increase. As λ approaches zero, $\frac{1}{\lambda^4}$ will approach infinity. Clearly, the experimental evidence did not agree with the classical physics theory. This failure of classical physics to predict the shape of the graph is often referred to as the **ultraviolet catastrophe**, because the irradiance of the blackbody radiation does not approach infinity in the uv region.

DON'T FORGET

Ultraviolet wavelengths are less than 400 nm.

Quantisation of black body radiation energy

In 1900 Max Planck offered an alternative explanation for the shape of the black body radiation graph which could explain the experimental evidence in the ultraviolet region. He suggested that the blackbody radiation energy emitted by hot objects only existed in

contd

integer multiples of hf, where h is a constant (now known as **Planck's constant**) and f is the wave frequency. The only values of blackbody radiation energies allowed were hf, $2hf$, $3hf$, $4hf$, and so on.

As the frequency increased so did the value of hf. Fractions of hf like $0 \cdot 5hf$ or $0 \cdot 1hf$ were not allowed. There would be no radiated energy for really high values of f (or small wavelengths) as hf is too big for any radiation to take place at this value of f. The term **quantum** was used to describe a quantity which could only exist in discrete integer values.

PHOTOELECTRIC EFFECT

Light falling on a metal plate can cause electrons to be emitted from the metal plate. This is the **photoelectric effect**. The diagram shows a light wave ejecting an electron from a zinc metal plate. The process is called **photoemission** and the electrons emitted are **photoelectrons**.

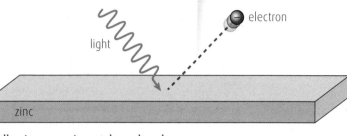

Classical physics wave theory could not explain the following experimental results when the metal used was zinc.

- A very bright intense visible light source did not cause an electron to be emitted from a clean zinc plate. Classical wave theory predicted that a bright source of light is a wave with a large amplitude and therefore has a large associated energy. No electron is ejected from the zinc plate regardless of how intense the visible light is.

- A very weak ultraviolet source is able to eject an electron from the surface of the zinc. Classical wave theory predicted there would be a time delay while the weak uv wave built up energy to eject the electron. There is no time delay observed.

Einstein's explanation

In 1905 Einstein took Planck's theory for blackbody radiation a stage further and proposed that all light is made up of discrete quanta of energy rather than continuous waves. The energy of each quantum of light is hf, agreeing with Planck. There is a **threshold frequency**, f_0, below which photoemission is not possible. The minimum energy required to cause photoemission from a metal is hf_0 and this minimum energy is called the **work function W** of the metal.

Einstein also predicted that the kinetic energy of a photoelectron ejected by light of frequency f is given by the relationship

$E_K = hf - W$

This expression can be represented graphically by

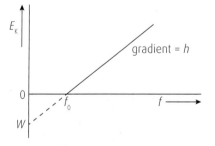

 DON'T FORGET

The formula $E_K = hf - W$ is like $y = mx + c$ and has a gradient h and intercept $-W$.

 **ONLINE**

Follow the link at www.brightredbooks.net for an online tutorial on the photoelectric effect.

THINGS TO DO AND THINK ABOUT

Most scientists were deeply sceptical of Planck's theory of blackbody radiation being quantised. Einstein kept an open mind and his 1905 theory of quantisation of all light waves was met with equal scepticism. In 1914 Robert Millikan set out to prove Einstein wrong but his experimental results on the photoelectric effect were not what he expected and they confirmed that light indeed was quantised. The term **photon** was not coined until 1926 by Gilbert Lewis.

It may be rewarding to carry out some internet research on the personalities involved in the early years of quantum theory.

 **ONLINE TEST**

Head to www.brightredbooks.net and test yourself on quantum theory.

QUANTUM THEORY: BOHR MODEL OF THE ATOM

BOHR MODEL OF THE ATOM

The model for the hydrogen atom suggested by Niels Bohr in 1913 had a **positive nucleus** (a proton), with an **electron** orbiting only in certain **allowed orbits**. The **angular momentum** of **each orbit** is **quantised**.

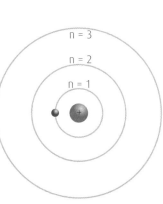

The **electron orbit circumference** had to contain a whole number (**n**) of de Broglie wavelengths.

Electrons can jump from one allowed orbit to another. The electron must gain energy when it jumps to an outer orbit. Energy is given off when the electron jumps from an outer orbit to an inner orbit.

The Bohr model is able to explain both emission line spectra and absorption line spectra.

hydrogen emission spectrum

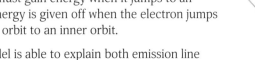

hydrogen absorption spectrum

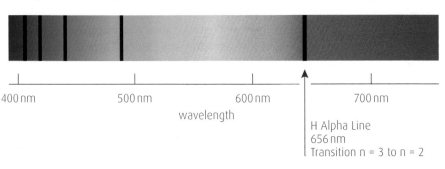

400 nm   500 nm   600 nm   700 nm

wavelength

H Alpha Line
656 nm
Transition n = 3 to n = 2

The red line on the emission spectrum has a wavelength of 656 nm. The Bohr model can explain this line as due to an electron jumping from orbit n3 to n2 releasing a photon of wavelength 656 nm. You should be able to show that the frequency of this photon is 4.57×10^{14} Hz and its energy is 3.03×10^{-19} J.

The dark line in the absorption spectrum of hydrogen at 656 nm is caused by an electron in the n2 orbit absorbing a photon of energy 3.03×10^{-19} J from an incident beam of white light. The electron jumps to orbit n3 before falling back to the n2 orbit releasing a photon of wavelength 656 nm. This released photon will not be in the same direction as the passing beam of white light. There will be a reduction in brightness in the passing white light corresponding to the wavelength 656 nm.

The other absorption lines can be explained by electrons moving between orbits different from n3 to n2.

QUANTISATION OF ANGULAR MOMENTUM

Bohr showed that the angular momentum of the orbiting electrons is also quantised. The angular momentum of an orbiting electron can only have values of $\frac{nh}{2\pi}$ where n is the orbit number. An electron in the first orbit ($n = 1$) has an angular momentum of $\frac{h}{2\pi}$. An electron in the second orbit ($n = 2$) has an angular momentum of $\frac{2h}{2\pi}$, and so on.

The angular momentum of the orbiting electron can also be given by the expression mrv where m is the mass of the electron, r is the orbit radius and v is the electron's speed.

Equating the two expressions for angular momentum gives the relationship

$$mrv = \frac{nh}{2\pi}$$

Example

a Calculate the angular momentum of an electron in the second orbit of a hydrogen atom.

b If the speed of the electron is $1{\cdot}1 \times 10^6\,ms^{-1}$ in the second orbit, calculate the radius of the second orbit.

Solution:

a $L = n\frac{h}{2\pi}$

$= 2 \times \frac{6{\cdot}63 \times 10^{-34}}{2 \times 3{\cdot}14}$

$= 2{\cdot}1 \times 10^{-34}\,kgm^2s^{-1}$

b $mvr = n\frac{h}{2\pi} = 2{\cdot}1 \times 10^{-34}$

$r = \frac{2{\cdot}1 \times 10^{-34}}{9{\cdot}11 \times 10^{-31} \times 1{\cdot}1 \times 10^6}$

$= 2{\cdot}1 \times 10^{-10}\,m$

⚙ EXERCISE

1 Calculate the smallest angular momentum that an electron can have in a hydrogen atom.

2 The angular momentum of an electron in the hydrogen atom is $4{\cdot}22 \times 10^{-34}\,kgm^2s^{-1}$. Which orbit is this electron in?

💭 THINGS TO DO AND THINK ABOUT

ONLINE TEST

Head to www.brightredbooks. net and test yourself on Bohr's model.

Classical physics suggests that an electron orbiting a nucleus will experience a centripetal force and will be accelerating towards the nucleus. Classical physics also predicts that any accelerating charged particle will radiate energy, so an orbiting electron should be radiating energy. This would make the electron spiral in towards the nucleus as its overall energy decreases. This does not happen and Bohr's model with fixed quantised orbits agreed with experimental observations at the time.

Nowadays, the Bohr model cannot explain some of the finer details of atomic spectra and has been succeeded by the **quantum mechanical model** of the atom.

QUANTUM THEORY: WAVE-PARTICLE DUALITY AND DE BROGLIE WAVELENGTH

WAVE-PARTICLE DUALITY

In the early 20th century, physicists had to come up with new theories to explain unexpected results found during some experiments. These new theories included:

- **electromagnetic radiation** having **particle** properties as well as **wave** properties
- **electrons** having **wave** properties as well as **particle** properties.

Electrons and electromagnetic radiation are said to exhibit **wave-particle duality**.

Evidence of wave-particle duality comes from these properties:

- electrons behaving as particles
- electron diffraction
- interference patterns.

Electrons as particles

An **electron** has **mass** and can be **accelerated**. These properties suggest that electrons are **particles**.

Electron diffraction

Electrons can be shown to **diffract** and produce **interference patterns**. A diffraction tube with a thin graphite film in front of the electron gun will exhibit electron diffraction as electrons pass through and interfere in the region beyond the electron gun.

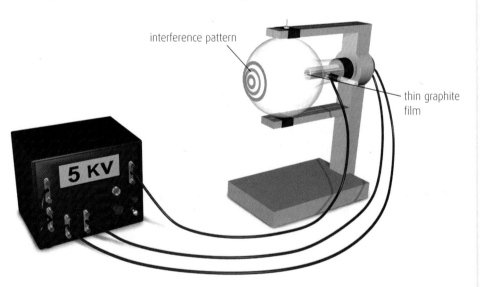

interference pattern

thin graphite film

5 KV

A series of concentric interference fringes will be seen on the fluorescent screen.

G.P. Thomson in Aberdeen was awarded the Nobel prize for his work demonstrating electrons producing an interference pattern. An interference pattern is the test for a wave.

Interference patterns in electromagnetic radiation

Electromagnetic radiation can produce **interference patterns**, as demonstrated with visible light passing through a double slit producing a series of interference fringes, as was seen in Higher Physics. This provides evidence that electromagnetic radiation has wave properties.

DON'T FORGET

Be careful with the spelling of the word 'diffract'. If you write 'defract' or 'deffract' in an exam you will lose marks as it is too close to the word 'refract'.

DON'T FORGET

Enter "wave-particle duality" into an internet search engine for more information.

DE BROGLIE WAVELENGTH

An expression for the wavelength of a photon can be derived.

The energy of a photon is

$$E = hf$$

$$= h \times \frac{c}{\lambda}$$

$$mc^2 = h \times \frac{c}{\lambda} \quad \text{substituting } E = mc^2$$

$$\lambda = \frac{h}{mc}$$

$$\lambda = \frac{h}{p} \quad \text{where } \boldsymbol{p} = \textbf{momentum (mass} \times \textbf{velocity)}$$

de Broglie proposed that this expression could be applied to both particles and waves.

Example

Calculate the wavelength of an electron travelling at $6.5 \times 10^6\,\text{ms}^{-1}$.

Solution:

$$\lambda = \frac{h}{p}$$

$$= \frac{h}{mv}$$

$$= \frac{6.63 \times 10^{-34}}{9.11 \times 10^{-31} \times 6.5 \times 10^6}$$

$$= 1.1 \times 10^{-10}\,\text{m}$$

A wavelength of $1.1 \times 10^{-10}\,\text{m}$ is comparable to the spacing between atoms, and so diffraction and interference effects are possible, as this wavelength is of the same order as the gaps between atoms.

Example

Calculate the wavelength of a bullet of mass $10\,\text{g}$ travelling at $120\,\text{ms}^{-1}$.

Solution:

$$\lambda = \frac{h}{p}$$

$$= \frac{h}{mv}$$

$$= \frac{6.63 \times 10^{-34}}{10 \times 10^{-3} \times 120}$$

$$= 5.5 \times 10^{-34}\,\text{m}$$

A wavelength of $5.5 \times 10^{-34}\,\text{m}$ is much smaller than any physical dimensions available, and so diffraction and interference effects associated with this bullet cannot be observed.

 DON'T FORGET

Relativistic effects can be ignored at speeds less than 10% of the speed of light.

EXERCISE

1   Calculate the wavelength of an electron travelling at 5% of the speed of light.

2   Electron A has twice as much kinetic energy as electron B. What is the ratio of de Broglie wavelengths of electron A to electron B?

Bohr model and de Broglie wavelength

Allowed electron orbits must have an exact number of de Broglie wavelengths fitting into the circumference. The fourth orbit (n = 4) must have four electron wavelengths fitting exactly into the orbit circumference as shown by the diagram.

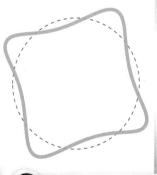

THINGS TO DO AND THINK ABOUT

The diagram above showing four electron wavelengths fitting exactly into the fourth orbit circumference is relatively easy to draw.

Try drawing the third circular orbit with three wavelengths fitting exactly – still relatively easy to draw.

Now try the second and first orbits with two and one wavelengths respectively fitting exactly on the circumferences – not so easy this time. An internet image search can be used to check your attempts.

 ONLINE TEST

Test yourself on this topic at www.brightredbooks.net

QUANTUM THEORY: UNCERTAINTY PRINCIPLE

WHAT IS THE UNCERTAINTY PRINCIPLE?

Classical physics views an electron as a small sphere of mass m and speed v. If the position and momentum of an electron are known, then the classical equations of motion can be used to predict exactly where the electron will be in the future.

In 1927, Werner Heisenberg questioned the validity of these precise predictions when dealing with particles in quantum physics. He reasoned that if the position of a quantum particle is known quite precisely, then its momentum will be less precisely known, and vice versa. **Heisenberg's uncertainty principle** states that it is impossible to determine accurately both the **position** and the **speed and direction** of a quantum particle **at the same instant**.

The following thought experiment illustrates the uncertainty principle. An electron with known momentum p passes through a wide slit of width Δx.

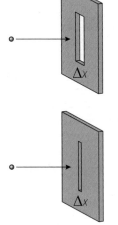

plan view

At the instant the electron passes through the slit its position is not precise. It can be anywhere across the width of the slit. The uncertainty in the position of the electron is large. The momentum p of the electron at the slit will be the same as it was before reaching the slit. The uncertainty in p at the slit is low.

Reducing the width of the slit will reduce the uncertainty in the position of the electron at the slit.

plan view

The narrow slit will increase diffraction effects and the electron has an increased range of possible directions after passing the slit. The direction (and momentum) of the electron will now be more uncertain.

It is important to note that these uncertainties are not a reflection on the quality of any measurements or measuring instruments. They are an inherent property of quantum physics.

Uncertainty relationship: position and momentum

The uncertainty in a quantum particle's position Δx and the uncertainty in the particle's momentum Δp are connected by the relationship

$$\Delta x \Delta p \geq \frac{h}{4\pi}$$

The product of these uncertainties must always be greater than or equal to $\frac{h}{4\pi}$. The numerical value of $\frac{h}{4\pi}$ is extremely small ($\approx 10^{-34}$). The product of the uncertainties only approaches $\frac{h}{4\pi}$ for very small particles.

Example

The uncertainty in an electron's position relative to an axis is given as $\pm 4\cdot 0 \times 10^{-12}$ m.

Calculate the least uncertainty in the simultaneous measurement of the electron's momentum relative to the same axis.

Solution:

$$\Delta x \Delta p \geq \frac{h}{4\pi}$$
$$\Delta p = \frac{h}{4\pi \Delta x} \qquad \text{for the least uncertainty in } \Delta p$$
$$= \frac{6\cdot 63 \times 10^{-34}}{4 \times 3\cdot 14 \times 4\cdot 0 \times 10^{-12}}$$
$$= \pm 1\cdot 3 \times 10^{-23} \, \text{kgms}^{-1}$$

contd

Some numerical questions may not give the momentum of a quantum particle but give a value of its speed with an associated uncertainty. You would have to find the mass of the particle from a data sheet then calculate the momentum using $p = mv$, as shown in the example.

Example

An electron moves along the x-axis with a speed of $4.75 \times 10^6\,ms^{-1} \pm 2\%$.

Calculate the minimum uncertainty with which you can simultaneously measure the position of the electron along the x-axis.

Solution:

momentum of the electron, $p = mv$

$$= (9.11 \times 10^{-31}) \times (4.75 \times 10^6)$$

$$= 4.327 \times 10^{-24}\,kgms^{-1}$$

$\Delta p = \pm 2\%$ of $4.327 \times 10^{-24}\,kgms^{-1}$ (assumes Δm is negligible]

$$= \pm 8.65 \times 10^{-26}\,kgms^{-1}$$

$\Delta x \Delta p \geq \frac{h}{4\pi}$

minimum uncertainty in position $\Delta x = \frac{h}{4\pi\Delta p}$

$$= \frac{6.63 \times 10^{-34}}{4 \times 3.14 \times 8.65 \times 10^{-26}}$$

$$= \pm\, 6.10 \times 10^{-10}\,m$$

Uncertainty relationship: energy and time

A similar relationship exists between the uncertainty in energy ΔE and the uncertainty in time Δt. The relationship is

$$\Delta E \Delta t \geq \frac{h}{4\pi}$$

If the time taken for an event is known quite precisely (small Δt) then the uncertainty in the energy associated with the event will be high (large ΔE)

Example

An electron spends approximately 3.5 ns in an excited state. Calculate the minimum uncertainty in the energy of the electron in this excited state.

Solution:

$\Delta E \Delta t \geq \frac{h}{4\pi}$

minimum uncertainty in the energy $\Delta E = \frac{h}{4\pi\Delta t}$

$$= \frac{6.63 \times 10^{-34}}{4 \times 3.14 \times 3.5 \times 10^{-9}}$$

$$= \pm 1.5 \times 10^{-26}\,J$$

Quantum mechanics

The study of the behaviour of quantum particles is called **quantum mechanics**. Predictions of what quantum particles will do is only possible in terms of probabilities and not certainties.

Quantum mechanics provides methods to determine these probabilities.

THINGS TO DO AND THINK ABOUT

When researching quantum mechanics you may come across the symbol $\hbar$ called **h bar**. This is a modified form of Planck's constant where $\hbar = \frac{h}{2\pi}$. It is used when angular frequency ω is used instead of frequency f so that the relationship $E = hf$ becomes $E = \hbar\omega$.

How would the relationship $\Delta x \Delta p \geq \frac{h}{4\pi}$ be rewritten to include $\hbar$ instead of h?

DON'T FORGET

Retain a greater number of sig. figs with intermediate calculations.

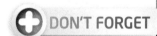

DON'T FORGET

minimum Δx uses = rather than $\geq$

DON'T FORGET

A large ΔE means it is large compared to Δt.

VIDEO LINK

Visit www.brightredbooks. net for a video presentation of Heisenberg's uncertainty principle.

ONLINE TEST

Test yourself on the uncertainty principle at www.brightredbooks.net

PARTICLES FROM SPACE: MOTION IN A MAGNETIC FIELD

You will see in the Electromagnetism Unit that there is a force on a conductor of length L carrying a current I in a magnetic field B. If the conductor is at an angle θ to the magnetic field, then the force F is given by the relationship $F = BIL\sin\theta$. It is reasonable to assume that the force was acting on the charges which made up the current in the conductor. Now we will look at the force that acts on a single charge which moves in a magnetic field.

RELATIONSHIP $F = Bqv$

Consider a **charge q** moving with a **uniform speed v** in a conductor of **length L** at right angles to a **magnetic field B**. The charge moves a distance L in **time t**.

Substitute $L = vt$ and $I = \frac{q}{t}$ into $F = BIL\sin\theta$

$F = BIL\sin\theta$

$\quad = B \times \left(\frac{q}{t}\right) \times (vt) \times \sin\theta$

$\quad = B \times q \times v \times \sin\theta$ consider $\theta = 90°$

$F = Bqv$

The force on a **single charge q** moving with **speed v** at **right angles** to a **magnetic field B** is **Bqv** even when the charge moves outside the confines of a conductor.

The direction of the force can be found by using the right-hand rule. The thumb will show the force direction on a negative charge (e.g. an electron).

For the force direction on a moving positive charge, use the right-hand rule as if the moving charge is an electron and then simply take the opposite direction as given by the thumb. The following examples will illustrate this.

Example

An electron moves with a speed of $7.5 \times 10^6\,\text{ms}^{-1}$ perpendicular to a magnetic field of magnetic induction 250 mT. Calculate the magnitude and direction of the force on the electron.

Solution:

$F = Bqv$

$\quad = (250 \times 10^{-3}) \times (1.6 \times 10^{-19}) \times (7.5 \times 10^6)$

$\quad = 3.0 \times 10^{-13}\,\text{N}$

Find the direction using the right-hand rule
- point second finger to the right (electron current)
- point first finger upwards (magnetic field)
- thumb points into page (force direction)

The force is **$3.0 \times 10^{-13}\,\text{N}$ into the page**.

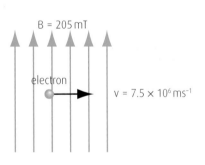

B = 205 mT

electron

v = 7.5 × 10⁶ ms⁻¹

Example

A proton moves with a speed of $1.8 \times 10^7\,\text{ms}^{-1}$ perpendicular to a magnetic field of magnetic induction 480 μT. The charge on a proton is $+1.6 \times 10^{-19}\,\text{C}$.
Calculate the magnitude and direction of the force on the proton.

Solution:

$F = Bqv$

$\quad = (480 \times 10^{-6}) \times (1.6 \times 10^{-19}) \times (1.8 \times 10^7)$

$\quad = 1.4 \times 10^{-15}\,\text{N}$

Find the direction using the right-hand rule
- point second finger down (ignore +charge at this stage)
- point first finger into page
- thumb points to the left.

An electron would move to the left.

The force on the proton will be **$1.4 \times 10^{-15}\,\text{N}$ to the right**.

× × × × × ×

× × × × × ×
proton

× × × | × × ×

× × × ▼ × × ×

× × × × × ×

B = 480 μT (into page)

DON'T FORGET

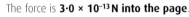

thrust (force)

magnetic field

electron current

The **first** finger represents the magnetic **field** direction.
The **second** finger represents the **electron** current direction.
The **thumb** represents the **thrust** (force).

DON'T FORGET

You will have met the right-hand rule in Higher Physics.

CIRCULAR PATH OF A CHARGED PARTICLE IN A MAGNETIC FIELD

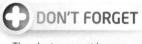

B (into page)

The **force** on a **charged particle moving at right angles** to a **magnetic field** is **perpendicular** to the **direction of travel**. This will cause the charged particle to change direction only, and its speed will be unaffected.

Consider an **electron** at **point A** in a **magnetic field** moving to the left with **velocity v**. The **force F** on the electron will be **upwards** and have a magnitude **Bqv**.

This causes the electron to **change direction** but not its **speed**. The force on the electron is a **central force**. At **point B**, the force will be **to the right**.

The electron moves **clockwise** in a circle under the influence of this central force.

A **positive charge** moving in the same magnetic field will also move in a circle and it will orbit **anticlockwise** as we look at it. The central force on the positive charge will act in the opposite direction to that of the negative charge (electron). The radius of the circular orbit will not necessarily be the same as for a negative charge.

DON'T FORGET

The electron must be considered to be moving in a vacuum to discount the possibility of collisions with air molecules.

Radius of orbit

We can now apply Newton's Second Law to find an expression for the **radius r** of the circular orbit of a charged particle moving perpendicularly to a magnetic field.

unbalanced force on charged particle = **mass × acceleration**

$Bqv = m \times \frac{v^2}{r}$ cancel v and rearrange

$r = \frac{mv}{Bq}$ where m = the mass of the charged particle.

The **radius r** is **directly proportional** to the **mass** and **velocity** of the charged particle and **inversely proportional** to the **magnetic induction** and **magnitude of the charge**.

Example

An electron of velocity $6\cdot8 \times 10^6$ ms^{-1} moves perpendicularly to a magnetic field of magnetic induction $0\cdot15$ T. Calculate the radius of its orbit.

Solution:

$r = \frac{mv}{Bq}$

$= \frac{9\cdot11 \times 10^{-31} \times 6\cdot8 \times 10^6}{0\cdot15 \times 1\cdot6 \times 10^{-19}}$

$= 2\cdot6 \times 10^{-4}$ m.

A proton with the same velocity as the electron moving in the same magnetic field will have a much bigger orbit radius, as the mass of the proton is bigger than the mass of the electron. (**B**, **q** and **v** would be unchanged.)

ONLINE

Head to www.brightredbooks.net to watch a video showing charges moving in a magnetic field.

Bubble chamber

The paths taken by charged particles in a magnetic field can be seen using a bubble chamber containing liquid hydrogen. The charged particles leave an ionisation track, and tiny bubbles form around the ions. These tracks can then be photographed for further study.

This photograph shows charged particles moving in a circle and spiralling inwards as they lose energy. (The radius decreases as the velocity decreases.) The particles can be identified using the magnitude and direction of the magnetic field as well as measurements of the initial radius.

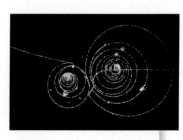

 THINGS TO DO AND THINK ABOUT

1   Practise using the right-hand rule. You will be expected to predict the direction of the force on a charged particle moving in a magnetic field.

2   The physics leading up to the relationship $r = \frac{mv}{Bq}$ is worth remembering.

ONLINE TEST

Head to www.brightredbooks.net and test yourself on motion in a magnetic field.

PARTICLES FROM SPACE: COSMIC RAYS

HELICAL MOTION IN A MAGNETIC FIELD

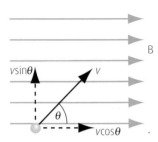

A charged particle entering a uniform magnetic field at 90° to the field lines will follow a circular path as previously described. If the particle enters the magnetic field at **angle** θ ($0° < \theta < 90°$), only a **component of the particle's velocity** will be **perpendicular** to the **magnetic field lines**.

The component of **v** perpendicular to **B** is $v_\perp = v\sin\theta$

The component $v_\perp$ will result in a force on the charged particle, causing it to change direction and move in a circle with a constant speed of $v\sin\theta$.

The component of **v** parallel to **B** is $v_\parallel = v\cos\theta$

There is no force on the charged particle parallel to the magnetic field, so $v_\parallel$ will remain constant.

The charged particle will move in a circle perpendicular to **B** as well as moving in a straight line parallel to **B**. This will result in the particle following a **helical path** as shown. The axis of the helix will be in the direction of the magnetic field.

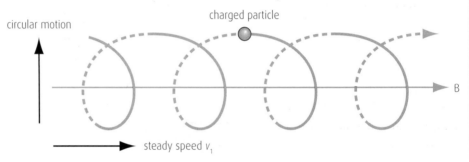

Looking from the left-hand side at the helical motion, you should see that the charged particle is moving clockwise. Use the right-hand rule to confirm that the charge on the particle must be negative. Positive charged particles would rotate anticlockwise as seen from the left.

Numerical calculations involving the helical path can be made (e.g. the radius of the helix), but care must be taken to use only the components of the particle's velocity and not the initial speed.

Example

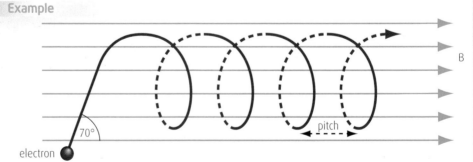

An electron travelling at $2{\cdot}3 \times 10^7\,\mathrm{ms^{-1}}$ enters a uniform magnetic field at an angle of 70° as shown. The magnetic induction = 0·18 T.

a Calculate the radius of the helix.

b Calculate the pitch of the helix (i.e. the distance between adjacent loops)

contd

Solution:

a
$$v_\perp = v\sin70 = 2\cdot3 \times 10^7 \times \sin70 = 2\cdot16 \times 10^7\,\text{ms}^{-1}$$

$$\text{Force on electron} = Bq(v\sin\theta) = 0\cdot18 \times (1\cdot6 \times 10^{-19}) \times (2\cdot16 \times 10^7)$$

$$= 6\cdot22 \times 10^{-13}\,\text{N}$$

$$F = m \times \frac{v^2}{r} \Rightarrow r = m \times \frac{v^2}{F} = \frac{9\cdot11 \times 10^{-31} \times (2\cdot16 \times 10^7)^2}{6\cdot22 \times 10^{-13}} = 6\cdot8 \times 10^{-4}\,\text{m}.$$

b $v_\parallel = v\cos70° = 2\cdot3 \times 10^7 \times \cos70° = 7\cdot87 \times 10^6\,\text{ms}^{-1}$

time for 1 revolution = $\frac{2\pi R}{v_\perp} = \frac{2 \times 3\cdot14 \times 6\cdot8 \times 10^{-4}}{2\cdot16 \times 10^7} = 1\cdot98 \times 10^{-10}\,\text{s}.$

pitch = $v_\parallel \times$ time or 1 revolution = $(7\cdot87 \times 10^6) \times (1\cdot98 \times 10^{-10}) = 1\cdot6 \times 10^{-3}\,\text{m}.$

Aurora Borealis (Northern Lights)

Charged particles emitted by the Sun and travelling towards the Earth will first enter the Earth's magnetic field. Unless θ is 0° or 90°, they will change direction and spiral along the Earth's magnetic field lines.

charged particle

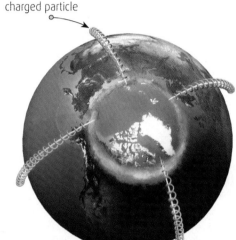

When the charged particles reach the Earth's atmosphere, they collide with air atoms and molecules producing light. Collisions with atomic oxygen produce green light while collisions with nitrogen produce pink light. A complete band of coloured light can be seen from space during periods of strong sunspot activity. From the Earth, an observer will see part of this band – the **Aurora Borealis** or **Northern Lights**.

This light display is regularly seen in the north of Scotland, and occasionally in the south of Scotland, well away from light pollution. Thanks go to Joshua Strang for the use of this photograph, taken in Alaska in 2005. Similar aurora can be seen near the South Pole, called the Aurora Australis.

THINGS TO DO AND THINK ABOUT

Cosmic rays are high energy charged particles originating in outer space. Most cosmic rays are nuclei of atoms travelling at close to the speed of light. Cosmic rays strike the Earth from all directions.

The **solar wind** is a stream of plasma released from the upper atmosphere of the Sun. It consists of mostly electrons, protons and alpha particles with energies usually between 1·5 and 10 keV.

Carry out an internet search to investigate further

Investigate the origin and composition of cosmic rays and the solar wind.

VIDEO LINK

Find out more about aurora by watching the clips at www.brightredbooks.net

ONLINE TEST

Test your knowledge of cosmic rays at www.brightredbooks.net

SIMPLE HARMONIC MOTION 1

INTRODUCTION TO SIMPLE HARMONIC MOTION

An oscillation or vibration describes the motion of an object which repeats a movement at regular time intervals. **Simple harmonic motion (SHM)** is a common type of oscillation where an object vibrates about an equilibrium position under the influence of an **unbalanced force** which is

1 always directed **towards** the **equilibrium** position

2 **proportional** to the object's **displacement** from the equilibrium position.

EQUATIONS FOR SHM

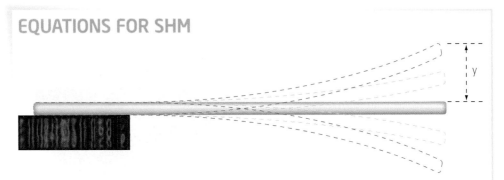

One end of a thin flexible rod vibrates vertically, performing SHM with **frequency f**. The **displacement y** of the end of the rod can be described by this expression using $\sin\omega t$.

$y = A\sin\omega t$ where A is the amplitude of the vibration, $\omega = 2\pi f$ and $y = 0$ when $t = 0$

Velocity is defined as the **rate of change of displacement**.

The velocity of the rod end is given by

$$v = \frac{dy}{dt}$$
$$= \frac{d}{dt}(A\sin\omega t)$$
$$= A\omega\cos\omega t$$

Acceleration is defined as the **rate of change of velocity**.

The acceleration of the rod end is given by

$$a = \frac{dv}{dt}$$
$$= \frac{d}{dt}(A\omega\cos\omega t)$$
$$= -A\omega^2\sin\omega t$$
$$= -\omega^2 y$$

acceleration $a = -\omega^2 y$

The maximum acceleration $= \pm\,\omega^2 A$ when $y = \pm A$

The velocity v of the rod end is

$$v = A\omega\cos\omega t$$
$$= \pm A\omega\sqrt{1 - \sin^2\omega t} \qquad \text{since } \sin^2\omega t + \cos^2\omega t = 1$$
$$= \pm A\omega\sqrt{1 - \frac{y^2}{A^2}}$$
$$v = \pm\,\omega\sqrt{A^2 - y^2}$$

The maximum velocity equals $\pm\omega A$ and occurs when $y = 0$.

contd

Example

The tip of one of the prongs of a tuning fork vibrates with SHM of frequency 440 Hz and amplitude 0·6 mm.

a Calculate the maximum acceleration of the tip.

b Calculate the maximum speed of the tip.

c Calculate **i** the acceleration

  ii the speed of the tip when the displacement of the tip is 0·25 mm from the equilibrium position.

Solution:

a First calculate ω.

$\omega = 2\pi f$

$= 2 \times 3\cdot14 \times 440$

$= 2\cdot76 \times 10^3\ \text{rads}^{-1}$

$a_{max} = \omega^2 A$

$= (2\cdot76 \times 10^3)^2 \times (0\cdot6 \times 10^{-3})$

$= 4\cdot6 \times 10^3\ \text{ms}^{-2}$

b $v_{max} = \omega A$

$= (2\cdot76 \times 10^3) \times (0\cdot6 \times 10^{-3})$

$= 1\cdot66\ \text{ms}^{-1}$

c i $a = -\omega^2 y$

$= -(2\cdot76 \times 10^3)^2 \times (0\cdot25 \times 10^{-3})$

$= -1\cdot9 \times 10^3\ \text{ms}^{-2}$ (the tip is decelerating when $y = (+)0\cdot25\,\text{mm}$)

ii $v = \pm\ \omega \sqrt{A^2 - y^2}$

$= 2\cdot76 \times 10^3 \times \sqrt{(0\cdot6 \times 10^{-3})^2 - (0\cdot25 \times 10^{-3})^2}$

$= 1\cdot5\ \text{ms}^{-1}$

EXERCISE

The displacement y of an object of mass 0·75 kg undergoing SHM is given by the equation

$y = 0\cdot25\sin45t$ metres

Calculate

a the frequency of the oscillation

b the maximum unbalanced force applied to the object

c the speed of the object at a displacement of 0·18 m.

THINGS TO DO AND THINK ABOUT

Not all repetitive periodic motion is SHM. The windscreen wipers on a car have a periodic motion with a certain frequency but move with steady speed for most of the time.

The condition for SHM requires the acceleration of the blades to be proportional to the displacement from a fixed point. The wiper blade motion is not SHM since the condition $a = -\omega^2 y$ is not met.

ONLINE TEST

Take the test on simple harmonic motion at www.brightredbooks.net

SIMPLE HARMONIC MOTION 2

OSCILLATING SPRINGS

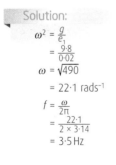

A spring hanging vertically has a **mass m** attached to its lower end. This causes the spring to stretch and increase in length. The increase in length of the spring, or **extension**, is labelled e_1 on the diagram.

The spring **extension** is **directly proportional** to the **tension** in the spring.

The **tension** in the spring equals **mg** when the forces are balanced (spring and mass stationary).

$mg \propto$ extension

$mg = k \times$ extension, where **k** is the **constant of proportionality**

In the above example $k = \dfrac{mg}{e_1}$

Now pull the mass down, increasing the length by a second extension e_2.

Release the mass and it will perform SHM because the unbalanced force is proportional to the extension (displacement) of the mass from its original position.

$$\text{Unbalanced force} = -\frac{mg}{e_1} \times \text{extension}$$

$$ma = -\frac{mg}{e_1} \times \text{extension}$$

$$a = -\frac{g}{e_1} \times \text{extension}$$

The amplitude of the SHM will be e_2 and $\omega^2 = \dfrac{g}{e_1}$

This derivation is extension material which is not examinable at this level.

DON'T FORGET ➕

Minus sign because the force and extension are in opposite directions.

Example

A mass of 0·6 kg causes an extension of 2 cm when hung on the end of a spring. The mass is pulled down a further 0·5 cm and released. Show that the frequency of oscillation is 3·5 Hz.

Solution:

$\omega^2 = \dfrac{g}{e_1}$

$\quad = \dfrac{9\cdot8}{0\cdot02}$

$\omega = \sqrt{490}$

$\quad = 22\cdot1 \text{ rads}^{-1}$

$f = \dfrac{\omega}{2\pi}$

$\quad = \dfrac{22\cdot1}{2 \times 3\cdot14}$

$\quad = 3\cdot5 \text{ Hz}$

SIMPLE PENDULUM

A popular class experiment in AH Physics is measuring **g** using a simple pendulum. The pendulum consists of a mass **m** on the end of a string of length **l** as shown.

The tangential component of the weight **mg**sinθ is the **unbalanced force** on the mass.

For small angles, $\sin\theta = \theta = \frac{arc\ length\ (x)}{l}$

The acceleration of the mass is

$$a = \frac{unbalanced\ force}{m}$$
$$= -\frac{mg\frac{x}{l}}{m}$$
$$= -\frac{g}{l}x$$

The motion must be SHM, as the acceleration is proportional to the displacement and $\omega^2 = \frac{g}{l}$.

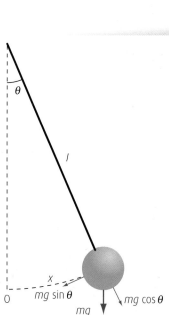

This leads to an expression for the **period T** of the pendulum.

$$T = \frac{2\pi}{\omega}$$
$$= 2\pi\sqrt{\frac{l}{g}} \qquad \text{or } T^2 = \frac{4\pi^2 l}{g}$$

The period **T** is measured for several different lengths **l** of the pendulum, and a graph of T^2 against **l** is drawn. The graph should be a **straight line through the origin** with gradient $\frac{4\pi^2}{g}$.

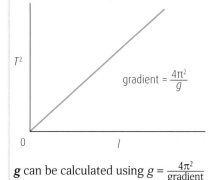

gradient = $\frac{4\pi^2}{g}$

g can be calculated using $g = \frac{4\pi^2}{gradient}$

DON'T FORGET

The minus sign is because displacement and unbalanced force are in opposite directions.

ONLINE

A pendulum with a very long string is called a **Foucault pendulum** and can show the rotation of the Earth. To see a video on this visit www.brightredbooks.net

THINGS TO DO AND THINK ABOUT

1 Use the following data to draw an appropriate graph to find a value of g. Note that the time of ten complete swings has been given.

length l (m)	0·50	0·70	0·90	1·2	1·5
10 periods (10T) (s)	14·3	17·0	19·2	22·2	24·8

2 Calculate the length of a pendulum whose period is 1 second. If you have time, try this experimentally and see if theory agrees with practice.

ONLINE TEST

Take the test on simple harmonic motion at www.brightredbooks.net

SIMPLE HARMONIC MOTION 3

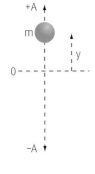

KINETIC AND POTENTIAL ENERGY IN SHM

As the mass on the end of a spring oscillates up and down, there is a constant **interchange** of **kinetic energy** to **potential energy** and vice versa. When the mass momentarily stops at the top and bottom of the oscillation, all the energy is potential energy. As the mass passes through the equilibrium position at maximum speed, all its energy is kinetic energy.

If no energy is lost to frictional forces, then the **total energy** ($E_k + E_p$) will be **constant**.

EXPRESSION FOR KINETIC ENERGY

A mass m oscillates in a straight line about **0** with an amplitude A. The motion is SHM. An expression for the kinetic energy E_k of mass m is derived as follows

$$E_k = \tfrac{1}{2}mv^2 = \tfrac{1}{2}m(\pm\omega\sqrt{A^2 - y^2})^2 = \tfrac{1}{2}m\omega^2(A^2 - y^2)$$

$$E_k = \tfrac{1}{2}m\omega^2(A^2 - y^2)$$

Note that when $y = 0$, the **kinetic energy** is **maximum**.

$$(E_k)_{max} = \tfrac{1}{2}m\omega^2 A^2$$

The total energy of the system E_{Total} will also equal $\tfrac{1}{2}m\omega^2 A^2$

When $y = A$, the **kinetic energy** is **zero**.

$$(E_k)_{min} = \tfrac{1}{2}m\omega^2(A^2 - y^2) = \tfrac{1}{2}m\omega^2(A^2 - A^2) = 0$$

EXPRESSION FOR POTENTIAL ENERGY

An expression for the **potential energy** E_p of a mass m undergoing SHM is derived.

$$(E_p)_{max} = (E_k)_{max} = \tfrac{1}{2}m\omega^2 A^2 \qquad \text{when } y = A$$

At displacement y

$$E_p = E_{Total} - E_k$$
$$= \tfrac{1}{2}m\omega^2 A^2 - \tfrac{1}{2}m\omega^2(A^2 - y^2)$$
$$= \tfrac{1}{2}m\omega^2 A^2 - \tfrac{1}{2}m\omega^2 A^2 - \left(-\tfrac{1}{2}m\omega^2 y^2\right)$$
$$E_p = \tfrac{1}{2}m\omega^2 y^2$$

The following graphs show how the kinetic and potential energies vary with displacement and time for a mass undergoing SHM.

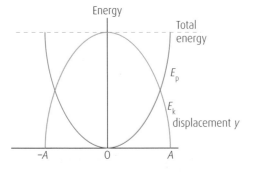

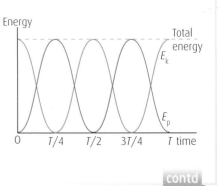

The mass has zero displacement at time $t = 0$.

contd

⚙ EXERCISE

1   A mass of 0·25 kg undergoes SHM with an amplitude of 0·48 m. If the total energy is 60 J, calculate the frequency of the oscillation.

2   The displacement of a mass undergoing SHM is exactly half the amplitude at one instant. Show that the kinetic energy of the mass is 75% of the total energy at this instant.

3   The graph shows how the potential energy of a 0·25 kg mass varies with displacement when performing SHM.

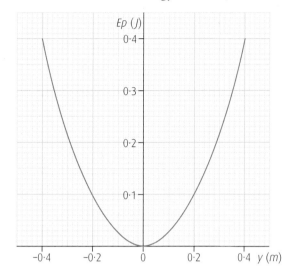

  a   What is the potential energy of the mass when the displacement is 0·2 m?

  b   What is the total energy of the mass?

  c   Calculate the kinetic energy of the mass when the displacement is 0·2 m.

  d   Calculate the speed of the mass when the displacement is 0·2 m.

  e   Calculate the maximum speed of the mass.

  f   What is the displacement when the speed is maximum?

DAMPING

The amplitude of oscillation of a pendulum will gradually decrease to zero due to air resistance on the moving parts of the pendulum. This is called **damping**. The total energy of the pendulum will decrease to zero. This energy will be transformed into heat energy given to the surroundings. The following graphs show how the displacement varies with time for undamped oscillations and for lightly damped oscillations.

Notice that the amplitude decreases with each half-damped oscillation.

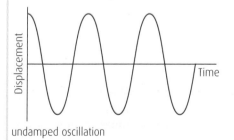

undamped oscillation

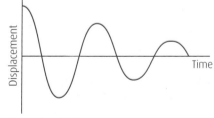

damped oscillation

➕ DON'T FORGET

It would be **incorrect** to say the **speed** decreases as a result of damping.

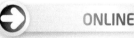

➡ ONLINE

To watch a video on energy in SHM visit www.brightredbooks.net

💭 THINGS TO DO AND THINK ABOUT

The **total** energy of an object undergoing SHM is either

- E_P (max) when $y = A$ or
- E_K (max) when $y = 0$

The formulae for E_P or E_K on the relationships sheet simply need to be modified.

✔ ONLINE TEST

Take the test on simple harmonic motion at www.brightredbooks.net

WAVES

A wave is a disturbance which moves from one place to another in a medium, **transferring energy** as it moves. Particles in the medium can be displaced as the wave passes but return to their undisturbed positions once the wave has passed through. There is **no net mass transport** in wave motion. Only energy is transferred.

The simplest mathematical model of a wave is based on the **sine** or **cosine** function. The diagram shows a **sine** wave which begins at **zero displacement**.

The **cosine** wave begins at **maximum positive displacement** on a crest.

PHASE DIFFERENCE

While you are already familiar with the terms **wavelength**, **frequency** and **amplitude** of a wave, the concept of **phase difference** is new.

Consider two points P and Q on a sine wave which are a distance x apart. The phase difference is a measure of the **separation** of these two points as a **fraction** of the wavelength and expressed as an **angle** in **radians**.

The phase difference $\Phi = \frac{x}{\lambda} \times 2\pi$

Points which are $\frac{1}{2}\lambda$ apart will have a phase difference of π.

Example

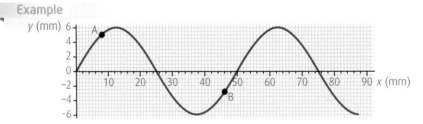

Calculate the phase difference between points A and B on the wave.

Solution:

$$\Phi = \frac{x}{\lambda} \times 2\pi$$
$$= \frac{46 - 8}{50} \times 2 \times 3{\cdot}14$$
$$= 4{\cdot}8\,\text{rad}$$

ENERGY TRANSFERRED BY A WAVE

As the **amplitude** of a wave **increases**, the **energy transferred** by the wave also **increases**. The energy E transferred by a wave is directly proportional to the square of the amplitude of the wave or $E = kA^2$. If the **amplitude** of a wave **doubles**, the **intensity** of the wave increases **four times**.

A useful formula for this direct-proportion relationship is

$$\frac{E_1}{(A_1)^2} = \frac{E_2}{(A_2)^2} \ (= \text{constant of proportionality})$$

DON'T FORGET ✚

This type of formula is useful for direct proportion, e.g. with the Gas Laws:

⚙ **EXERCISE**

1 How does the amplitude change when the energy transferred by a wave doubles?

2 How does the energy transferred by a wave change when the amplitude halves?

EQUATION OF A TRAVELLING WAVE

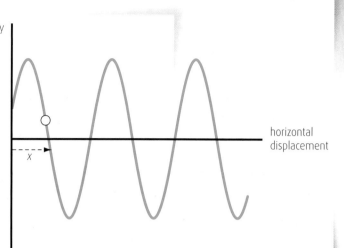

A travelling wave can be generated on an elasticated rope by repeatedly moving one of its ends up and down.

The wave moves to the right, and each point on the rope changes position with time. The diagram shows the profile of the wave at one instant in time.

horizontal displacement

The displacement y of a point on the rope located at the origin is given by the expression

$y = A\sin\omega t$

$\quad = A\sin2\pi ft$ where f is the wave **frequency** and A is the wave **amplitude**.

At a horizontal distance $+x$ from the origin, the displacement of a rope particle is given by

$y = A\sin\left(2\pi ft - \frac{x}{\lambda} \times 2\pi\right)$

$y = A\sin2\pi\left(ft - \frac{x}{\lambda}\right)$

The equation of an identical wave travelling to the left is given by

$y = A\sin2\pi\left(ft + \frac{x}{\lambda}\right)$

DON'T FORGET

Moving $\sin\theta$ to the right by **a** becomes $\sin(\theta - \textbf{a})$. You should be familiar with this from your studies of maths.

Example

A travelling wave has the equation $y = 0.65\sin(5t - 3x)$ where x and y are in metres and t in seconds. Determine the amplitude; frequency; wavelength and speed of the wave.

Solution:

By inspection, the amplitude = 0·65 m.

By calculation

$2\pi f = 5$ (comparing coefficients of t)

$\quad f = \frac{5}{2 \times 3.14} = 1.27\,Hz$

Also

$-\frac{2\pi}{\lambda} = -3$

$\quad \lambda = \frac{2 \times \pi}{3} = 2.1\,m.$

The speed of the wave is calculated.

$v = f \times \lambda = 1.27 \times 2.1 = 2.7\,ms^{-1}.$

⚙ EXERCISE

3   The equation of a wave is $y = 2.5 \times 10^{-3} \times \sin(45t + 0.7x)$. x, y in m and t in s. Calculate the amplitude, frequency, wavelength, speed and direction of the wave.

4   Write down the equation of a wave travelling to the right with a speed of $0.75\,ms^{-1}$, wavelength 2·5 cm and amplitude of 4 cm.

➡ ONLINE

Visit www.brightredbooks.net for an online tutorial on the wave equation.

💭 THINGS TO DO AND THINK ABOUT

The equation of a travelling wave $y = A\sin2\pi\left(ft - \frac{x}{\lambda}\right)$ can be rewritten to include wave speed v and period T.

$y = A\sin2\pi\left(\frac{t}{T} - \frac{x}{\lambda}\right) y = A\sin2\pi\left(\frac{t}{T} - \frac{x}{\lambda}\right)$

$y = A\sin2\pi f\left(t - \frac{x}{v}\right)$

Show that these three equations are equivalent.

The sine of an angle has no units or dimensions. Show that this is the case with each of these three expressions for a travelling wave.

✓ ONLINE TEST

Test your knowledge of waves at www.brightredbooks.net

STATIONARY WAVES

STATIONARY WAVES: OVERVIEW

A travelling wave moving to the right along a stretched slinky can be reflected if the other end of the slinky is fixed.

The incident wave and the reflected wave will interfere, and a stationary wave is set up at various wave frequencies.

One such stationary wave looks like this

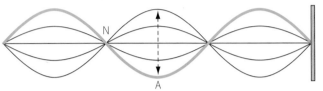

Some points on the stationary wave do not move. These points are called **nodes** (*N*). Nodes are positions of **zero disturbance** in a stationary wave. The diagram shows four nodes. **Halfway** between the nodes are points of **maximum disturbance** called **antinodes** (*A*). This diagram shows three antinodes.

Note also that the **distance between nodes** equals **half a wavelength**.

MICROWAVE STATIONARY WAVES

Microwaves are transmitted towards a metal plate which reflects the waves back towards the transmitter. A stationary wave is produced between the incident and reflected waves. The location of nodes can be found using a microwave detector connected to an ammeter.

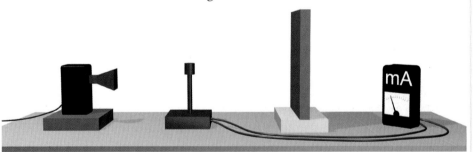

The ammeter will show a **minimum** reading when the detector is at a **node**. Note the positions of two adjacent nodes. The distance between two adjacent nodes is equal to $\frac{1}{2}\lambda$.

Typical results

distance between 7 consecutive nodes = 8·4 cm
distance between 2 consecutive nodes = 1·4 cm

$$\frac{\lambda}{2} = 1\cdot4$$

$$\lambda = 2\cdot8\,\text{cm}$$

The frequency of the microwave transmitter can now be calculated and compared to the value given by the manufacturer.

$$v = f \times \lambda$$

$$f = \frac{v}{\lambda} = \frac{3 \times 10^8}{2\cdot8 \times 10^{-2}} = 1\cdot07 \times 10^{10}\,\text{Hz}$$

STATIONARY SOUND WAVES

Sound from a loudspeaker can be sent along a closed tube, and the **reflected** sound wave will **interfere** with the **incident** wave to set up a **stationary** wave at various **frequencies** or **lengths of tube**. The frequency of the sound is adjusted using the dial on the signal generator until a loud resonant sound is heard coming from the tube. A **stationary sound wave** has been formed in the tube. The stationary sound wave is often represented as a **transverse** wave with **nodes** (*N*) and **antinodes** (*A*). Notice that a **node is formed** at the **closed** end of the tube.

One method of locating the nodes is to place some fine powder such as cork dust, along the base of the tube.

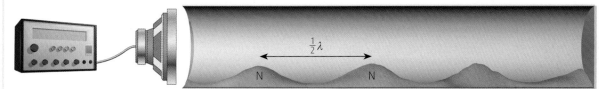

Switching on the signal generator and adjusting the frequency until a resonant sound is heard causes the layer of powder to change shape.

The powder will experience forces due to the disturbance of the air molecules at and near the antinodes. Powder at the nodes will not be disturbed, and the amount of powder here will build up.

⚙ EXERCISE

1   Calculate the speed of sound using the following results.
    distance between adjacent nodes = 92 mm
    sound frequency = 1750 Hz

2   Small mounds of powder are formed 12 cm apart in the tube above. If the speed of sound is 340 ms^{-1}, calculate the frequency of the sound.

VIDEO LINK

Visit www.brightredbooks.net for video and explanation of Kundt's tube.

☁ THINGS TO DO AND THINK ABOUT

ONLINE TEST

Test your knowledge of waves at www.brightredbooks.net

The standing wave is caused by the **superposition** (or overlapping) of the incident wave and the reflected wave.

Mathematically, the displacement *y* will be

$$y = A\sin 2\pi\left(ft - \frac{x}{\lambda}\right) + A\sin 2\pi\left(ft - \frac{x}{\lambda}\right)$$

$$= 2A\sin(2\pi ft)\cos\left(2\pi \frac{x}{\lambda}\right) \text{ using a trig addition relationship.}$$

When $x = 0, \frac{1}{2}\lambda, \lambda, \frac{3}{2}\lambda, \ldots$ the value of y is a maximum since $\cos\left(2\pi \frac{x}{\lambda}\right)$ will be ±1. Antinodes occur every half-wavelength. Nodes which are midway between antinodes also occur every half wavelength.

Note that the amplitude of the standing wave is $2A\cos\left(2\pi \frac{x}{\lambda}\right)$ and is dependent on the value of *x*.

INTERFERENCE: DIVISION OF AMPLITUDE

COHERENCE

Two travelling waves are said to be **coherent** if there is a **constant phase difference** between them.

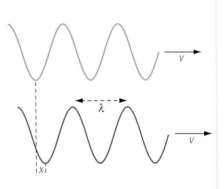

The constant phase difference Φ between these coherent waves is

$$\Phi = \frac{x}{\lambda} \times 2\pi$$

Destructive interference occurs if the **phase difference** between two overlapping coherent waves is π. This happens when $x = \frac{1}{2}\lambda$ and is equivalent to a **crest superimposed on a trough**.

DON'T FORGET

Explaining destructive interference in terms of crest and trough is very basic. Phase-difference explanations are more appropriate at AH level.

Two loudspeakers in parallel will produce coherent sound waves, as each speaker cone will move exactly in step with the other. Producing coherent light waves using two lamps in parallel is not possible due to the random nature of electron transitions and photon production in each lamp filament. Methods of producing coherent light waves will be described shortly.

Optical path difference

Interference in Higher Physics is explained in terms of path difference between two interfering waves. The concept of path difference must now be refined to include one of the waves passing through a medium other than air.

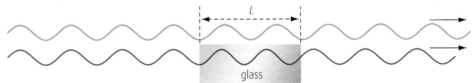

The diagram represents two rays of light of equal geometric length, one travelling in **air** and the other passing through a **block of glass** of length **L**.

DON'T FORGET

$\lambda_{air} = n\lambda_{medium}$

The **wavelength** in the glass **decreases** due to **refraction** and emerges out of phase with the wave above. It can be shown that a length **L** in a medium of refractive index **n** is equivalent to a length **nL** in air.

The **geometric** path length in the glass $= L$
The **optical** path length in the glass $= n_{glass}L$
The optical path difference (OPD) between the two rays $= n_{glass}L - n_{air}L$

DON'T FORGET

$n_{air} = 1$

$$OPD = n_{glass}L - L$$

Consider now two rays of light at near-normal incidence reflected off a glass block of refractive index n. One ray reflects off the front surface of the block while the other reflects internally off the rear surface. The glass block has a length 6 cm and refractive index 1·5.

geometric path length in the glass = 2 × 6 cm = 12 cm
(there and back)

optical path length in the glass = n × geometric path length

The top ray doesn't travel this extra distance.

Optical path difference between the two rays = 1·5 × (12 × 10⁻²)

$= 0·18$ m

contd

Phase changes on reflection

A light wave in air **reflecting off glass** undergoes a phase change of π.

A **crest** reflects as a **trough**.

The change of phase can be seen by reflecting a single slinky pulse from a fixed end. The pulse is reflected the other way round.

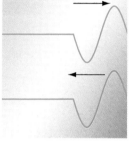

A **phase change of** π occurs when light is reflected from a **higher refractive index** medium (or **optically denser** medium).

A **light wave in glass** reflecting back **into the glass** at a glass/air boundary has **no change in phase**.

A **crest** reflects as a **crest**.

There is **no phase change** at this reflection.

There is **no phase change** when light reflects from an interface with a **lower optical density** medium (**decrease in refractive index**).

Division of amplitude

Coherent sources of light are required to produce observable interference fringes.

One way of doing this is to take one light wave and split it into two waves by reflection and refraction, then recombine these two waves later.

Notice that the **amplitude** of the incident wave is **greater** than the amplitudes of the **reflected** and **transmitted** waves.

An interference pattern produced by this method is called **interference by division of amplitude**. Extended light sources like the Sun or fluorescent lights can be used to produce visible interference patterns using division of amplitude.

transmission
(refraction)

reflection

THINGS TO DO AND THINK ABOUT

Incoherent waves also interfere but do so in a random manner. Destructive interference can only last for an instant with incoherent waves. Think of a busy swimming pool where all the water waves have different wavelengths and are out of phase with each other. No regular interference pattern is seen.

Coherent waves, on the other hand, set up regular interference patterns which stay constant with time and are mathematically predictable.

A laser is one light source that does produce coherent light because each photon is emitted precisely in phase with the stimulating photon.

If the OPD (optical path difference) is a whole number of wavelengths, the waves will be in phase ($\Phi = 0$).

If the OPD equals $\frac{1}{2}\lambda$, or $\frac{3}{2}\lambda$, or $\frac{5}{2}\lambda$, ... the waves will be completely out of phase ($\Phi = \pi$).

VIDEO LINK

Visit www.brightredbooks.net for a good video on phase changes on reflection.

ONLINE TEST

Head to www.brightredbooks.net and test yourself on interference.

INTERFERENCE: OIL FILMS

OIL FILMS: OVERVIEW

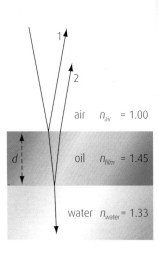

A thin film of oil on a puddle of water appears multicoloured when viewed in daylight. This is an example of interference by division of amplitude.

To understand what is happening, we will consider a ray of monochromatic light falling almost vertically on the oil.

The refractive index of oil is 1·45, while water has a refractive index of 1·33.

The ray of monochromatic light is partially reflected and partially transmitted.

The transmitted ray reflects off the bottom surface of the oil and emerges back into the air.

It is these two **reflected** rays which interfere and cause the oil-film colours.

Ray 1 reflects with a phase change of π ($n_{oil} > n_{air}$). A phase change of π is equivalent to $\frac{1}{2}\lambda$.

Ray 2 has passed through the oil-film thickness and reflected back with no change of phase ($n_{water} < n_{oil}$).

The optical path difference (OPD) between rays 1 and 2 = $2\,n_{oil}d + \frac{\lambda}{2}$

For destructive interference, the OPD must equal $\frac{\lambda}{2}$ or $1\frac{1}{2}\lambda$ or $2\frac{1}{2}\lambda$ etc. (i.e. an odd number of half-wavelengths).

Combining these gives $2n_{oil}d + \frac{\lambda}{2} = \left(m + \frac{1}{2}\right)\lambda$ giving $2n_{oil}d = m\lambda$ for destructive interference.

Example

Calculate the minimum thickness of oil which will result in destructive interference of red light ($\lambda = 650\,nm$) at near-normal incidence ($n_{oil} = 1\cdot45$).

Solution:

$2n_{oil}d = m\lambda$

$2 \times 1\cdot45 \times d = 1 \times 650 \times 10^{-9}$

($m = 1$ will give the minimum value for d)

$d = 2\cdot24 \times 10^{-7}\,m\ (0\cdot22\,\mu m)$

Removing this wavelength from white light will cause the reflected light from the oil on the puddle to appear coloured (green/blue). In practice, the oil film on the puddle will have many different thicknesses, so many reflected wavelengths will interfere destructively. Removing yellow light from white light will give the reflected light a purple hue.

Constructive interference is also possible between the two reflected rays. This happens when the OPD is equal to a whole number of wavelengths.

$OPD = 2\,(n_{oil})d + \frac{\lambda}{2} = m\lambda$, where $m = 1, 2, 3, \ldots$

ONLINE

Visit www.brightredbooks. net for explanation of colours on a CD and more.

EXERCISE

1 The oil film in the example has a thickness of $2\cdot24 \times 10^{-7}\,m$. Show that this film will permit constructive interference of wavelengths 1300 nm (IR) and 433 nm (violet).

BLOOMED LENSES

One practical application of thin-film interference is anti-reflection coatings on camera and binocular lenses. A thin coating of magnesium fluoride is applied to the front of the lens during manufacture. A light wave reflects from the front and rear of this coating – and if these two rays interfere destructively, then the light wave will not be reflected. The resultant picture quality is improved as more light is transmitted to the digital sensors or film.

contd

In more detail, consider monochromatic light falling on the coated lens at near-normal incidence.

Ray 1 has a phase change of π on reflection because $n_{coating} > n_{air}$.

Ray 2 **also** has a phase change of π on reflection because $n_{glass} > n_{coating}$.

Optical path difference between rays 1 and 2 = $2(n_{coating})d$

For destructive interference, OPD must be $\left(m + \frac{1}{2}\lambda\right)$ with $m = 0$

(Lens coatings are applied as thinly as possible, hence $m = 0$)

$OPD = 2(n_{coating})d = \frac{1}{2}\lambda$

$d = \frac{\lambda}{4n}$

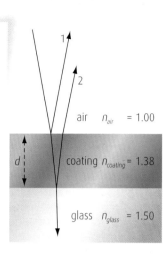

air n_{air} = 1.00

d coating $n_{coating}$ = 1.38

glass n_{glass} = 1.50

Example

Calculate the minimum thickness of magnesium fluoride that minimises reflection of light of wavelength 570 nm.

Solution:

$d = \frac{\lambda}{4n}$

$= \frac{570 \times 10^{-9}}{4 \times 1 \cdot 38}$

$= 8 \cdot 5 \times 10^{-8} \, m$

> **DON'T FORGET**
>
> n is the refractive index and does not have any identifying subscript on the data sheet.

There will be very little yellow light of wavelength 570 nm reflected by this thickness of coating on a glass lens. There will be some reflection of colours on either side of 570 nm in the visible spectrum. The relatively greater amounts of red and blue reflected light combine to give the lens a purplish hue.

| 700 nm | 650 nm | 600 nm | 550 nm | 500 nm | 450 nm | 400 nm |

| Red | Orange | Yellow | Green | Blue | Indigo | Violet |

relative strength
of reflected light

EXERCISE

2   Calculate the minimum thickness of magnesium fluoride that minimises light of frequency $5 \cdot 5 \times 10^{14}$ Hz.

3   A lens has a magnesium fluoride coating of thickness $1 \cdot 03 \times 10^{-7}$ m. Calculate the wavelength of light for which this lens is non-reflecting.

THINGS TO DO AND THINK ABOUT

Some binoculars and sunglasses have a coating with a ruby red appearance. This coating is designed to deliberately **reflect** a wavelength of this colour, and they look cool with their ruby tint. The downside is that the transmitted red colours can look a bit washed out, particularly when used on binocular lenses.

Top-end lenses have multiple layer coatings, and some internet research on this provides good background reading.

The effect of thin films improving light transmission was first observed by Lord Rayleigh in 1886, who noticed, by chance, that older, tarnished lenses gave 'better' light transmission than new lenses.

> **ONLINE TEST**
>
> Head to
> www.brightredbooks.net
> and test yourself
> on interference.

INTERFERENCE: THIN WEDGE INTERFERENCE

THIN WEDGE INTERFERENCE: AN OVERVIEW

A thin wedge of air between two glass plates will produce interference fringes by division of amplitude.

The incident wave is reflected from the bottom surface of the top glass plate and the top surface of the lower glass plate. The path difference between rays 1 and 2 is $2t$.

Ray 2 undergoes a phase change of π, as it reflects off glass in air ($n_{glass} > n_{air}$).

Ray 1 has no phase change on reflection, as it reflects off air inside glass ($n_{air} < n_{glass}$).

The optical path difference between rays 1 and 2 = $2t + \frac{\lambda}{2}$

For destructive interference, the usual condition is

$OPD = \left(m + \frac{1}{2}\right)\lambda$

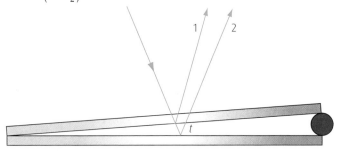

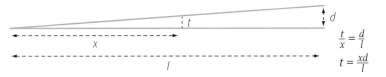

$$\frac{t}{x} = \frac{d}{l}$$

$$t = \frac{xd}{l}$$

Combining these two expressions gives

$2t + \frac{\lambda}{2} = \left(m + \frac{1}{2}\right)\lambda$

$2t = m\lambda \qquad$ substitute an expression for t found using similar triangles

$2\frac{xd}{l} = m\lambda$

$x = \frac{m\lambda l}{2d} \qquad$ This is the distance x of a minimum (dark) fringe from the apex of the wedge.

As x increases, the air gap t between the two plates also increases. The next minimum will occur at x' when $2t$ has increased by one wavelength.

$x' = \frac{(m+1)\lambda l}{2d}$

$\Delta x = x' - x \qquad$ Δx is the distance between two adjacent minima.

Δx is also the distance between two maxima, or fringe separation.

$= \frac{(m+1)\lambda l}{2d} - \frac{m\lambda l}{2d}$

$\Delta x = \frac{\lambda l}{2d}$

l is the length of the glass plate and d is the width of the open end of the wedge. d could be the thickness of a sheet of paper or the diameter of human hair. Wedge interference provides a method of measuring the dimensions of very thin objects.

DON'T FORGET

This derivation is not examinable although wedge fringes are in the syllabus.

DON'T FORGET

You will not be expected to derive this expression but the physics and maths involved should be quite accessible.

MEASURING THE DIAMETER OF A THIN WIRE

An air wedge can be formed by placing a thin wire between two microscope slides.

A sodium light source provides the monochromatic light. A glass beam splitter reflects this light vertically down onto the air wedge.

A travelling microscope is used to view the fringes and measure their separation.

Typical results

10 fringe widths = $6{\cdot}5 \times 10^{-4}$ m

length of glass plate = 80 mm

wavelength of light = 589 nm

$\Delta x = \frac{6{\cdot}5 \times 10^{-4}}{10} = 6{\cdot}5 \times 10^{-5}$ m

$\Delta x = \frac{\lambda l}{2d}$

$d = \frac{\lambda l}{2\Delta x}$

$\quad = \frac{(589 \times 10^{-9}) \times (80 \times 10^{-3})}{2 \times (6{\cdot}5 \times 10^{-5})}$

$\quad = 3{\cdot}6 \times 10^{-4}$ m

microscope view

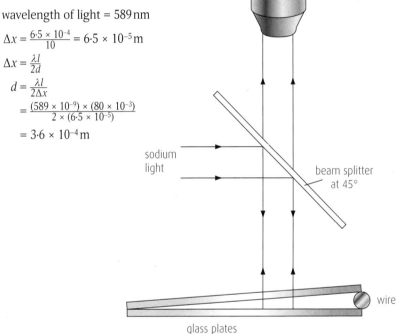

cross wires

travelling microscope

sodium light

beam splitter at 45°

wire

glass plates

➕ **DON'T FORGET**

Δx is found by dividing by 10 and **not** 9. 10 fringe widths is the same as the distance between 11 fringes.

⚙ EXERCISE

The wire in the experiment above is replaced by a sheet of paper of thickness $1{\cdot}7 \times 10^{-4}$ m.

Calculate the fringe separation.

➡ **ONLINE**

Visit www.brightredbooks.net for a video presentation on wedge fringes.

💭 THINGS TO DO AND THINK ABOUT

If water replaced air in the wedge, the fringe separation would be $\Delta x = \frac{\lambda l}{2nd}$ (n = refractive index of water). Try writing out the steps to prove this relationship.

What would happen to the fringe separation if each of the following is increased?

a   wavelength

b   refractive index

c   the angle of the wedge (try substituting $\tan\theta$ into the relationship).

✓ **ONLINE TEST**

Head to www.brightredbooks.net and test yourself on interference.

INTERFERENCE: DIVISION OF WAVELENGTH

WAVEFRONTS

Circular waves spread out radially and are usually drawn as a series of concentric circles representing wavefronts travelling out in all directions.

Each wavefront joins points on the wave which are in phase and have the same wavelength and frequency. Two points on the same wavefront can become coherent sources, which can lead to an interference pattern.

This applies equally to straight plane waves. All points on the same wavefront are potential sources of coherent light.

Interference by division of wavefront takes place between two **coherent waves** which originated from the **same wavefront**. An example of interference by division of wavefront is **Young's double slit experiment**, which was studied in Higher Physics.

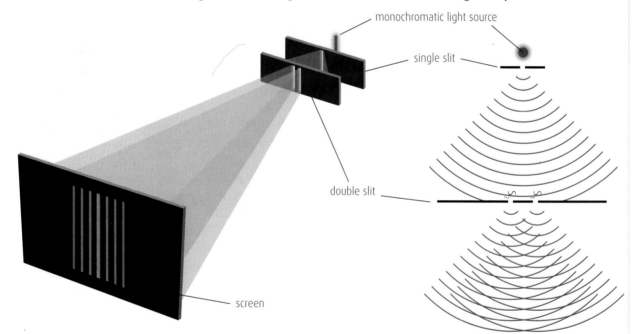

monochromatic light source

single slit

double slit

screen

The single slit, or **collimator**, ensures that a point source of coherent wavefronts moves towards the double slit. The double slit then becomes the source of two coherent waves which diffract and interfere, producing the interference fringes on the screen.

DON'T FORGET

The syllabus requires an explanation of Young's slits interference. The complete derivation however is not required.

YOUNG'S SLITS INTERFERENCE THEORY

Derivation of $\Delta x = \dfrac{\lambda D}{d}$

Consider the region between the double slit and the screen.

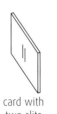

card with two slits

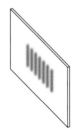

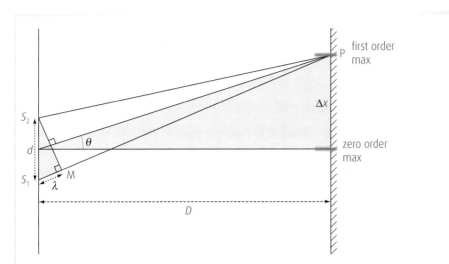

The second diagram shows a plan of the region between the zero-order and first-order maxima.

S_1 and S_2 represent the two slits.

The path difference $S_1P - S_2P = S_1M = \lambda$ as P is the first maximum.

The two shaded triangles are similar, so comparing similar sides gives

$\frac{\Delta x}{D} = \frac{\lambda}{d}$

$\Delta x = \frac{\lambda D}{d}$

The derivation assumes that angle θ is small, and so $S_2M \cong S_1S_2$.

Use your calculator to confirm that $\sin\theta \cong \tan\theta$ for small values of θ.

ONLINE

Measuring the wavelength of laser light using Young's slits apparatus is a standard physics experiment. Carry out an internet search for the details of this experiment.

Example

Monochromatic light incident on a double slit with a slit separation of 5.7×10^{-5} m gives this interference pattern on a screen 2·5 m away from the slits. Calculate the wavelength of the light.

80 mm

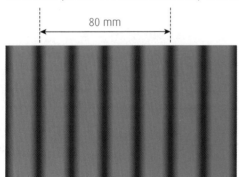

Solution:

4 fringe widths $= 80 \times 10^{-3}$ m

$\Delta x = 20 \times 10^{-3}$ m

$\Delta x = \frac{\lambda D}{d}$

$\lambda = \frac{\Delta x \times d}{D}$

$= \frac{(20 \times 10^{-3}) \times (5.7 \times 10^{-5})}{2.5}$

$= 456 \times 10^{-9}$ m (456 nm)

THINGS TO DO AND THINK ABOUT

Visit http://surendranath.tripod.com/Applets/Optics/Slits/DoubleSlit/DblSltApplet.html for an excellent interactive simulation of Young's slits.

ONLINE TEST

Head to www.brightredbooks. net and test yourself on interference.

POLARISATION

POLARISATION: OVERVIEW

Unpolarised light has oscillations (of the light's electric field) in **every** plane perpendicular to its direction of travel.

Polarising material (**polariser**) allows the transmission of light in one plane only due to the molecular alignment inside the material. The diagram shows transmission in the vertical plane only. (Some faint lines have been added to the polariser for purely explanatory purposes to indicate the plane of transmission.)

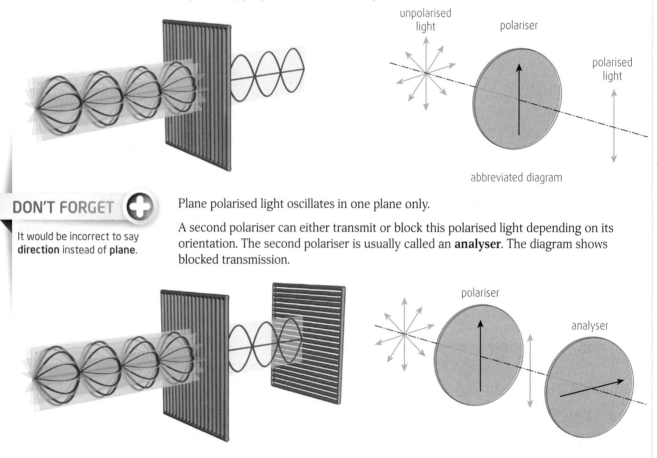

abbreviated diagram

Plane polarised light oscillates in one plane only.

A second polariser can either transmit or block this polarised light depending on its orientation. The second polariser is usually called an **analyser**. The diagram shows blocked transmission.

As the analyser is rotated slowly, transmission of the polarised light will increase and reach a maximum when the angle of rotation reaches 90°.

Only transverse waves can be polarised. Sound waves cannot be polarised, as they are not transverse.

POLARISATION BY REFLECTION

Unpolarised light reflected off the surface of an electrical insulator like glass or water can have its reflected light fully polarised. When this happens, the angle of incidence (and reflection) is called **Brewster's angle** (i_P), and the angle between the reflected (polarised) ray and the refracted ray is 90°. It can be shown that $\tan i_P = n$ where n is the refractive index of the reflecting material.

DERIVATION OF TAN $i_P = N$

The incident ray reflects at angle i_p and refracts at angle r. The angle between the reflected and refracted rays is 90°.

The refractive index n is

$$n = \frac{\sin i_p}{\sin r}$$
$$= \frac{\sin i_p}{\sin(180 - i_p - 90)}$$
$$= \frac{\sin i_p}{\sin(90 - i_p)}$$
$$= \frac{\sin i_p}{\cos i_p}$$
$$= \tan i_p$$

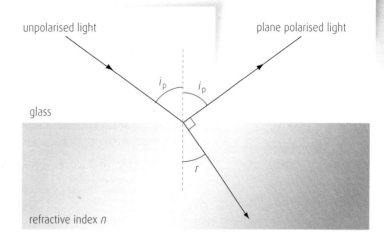

unpolarised light   plane polarised light

glass

refractive index n

EXERCISE

1 At what angle of incidence will unpolarised light reflect off water so that the reflected light is fully polarised? The refractive index of water is 1·33.

2 Unpolarised light reflects fully polarised off glass when the angle of reflection is 57°. Calculate the refractive index of the glass.

POLARISING SUNGLASSES

Sunlight, reflected off water at Brewster's angle, is plane polarised in the horizontal plane. Polarising sunglasses only transmit light oscillating in the vertical plane, so light reflected off water will not pass through.

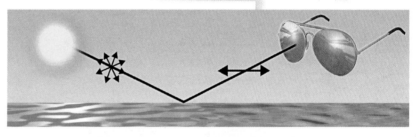

Reflections off the water, or glare, are removed, and the viewer is able to see more detail on and under the water. Photographs of water benefit from fitting a polarising filter to the camera lens.

Using polarising filter   No filter

ONLINE

Visit
www.brightredbooks.net
for a video demonstrating
polarisation effects.

ONLINE TEST

Test your knowledge
of polarisation at
www.brightredbooks.net

THINGS TO DO AND THINK ABOUT

1 The latest 3D cinema screens use vertically and horizontally polarised light. The spectacles worn in 3D cinemas have polarising lenses, and each eye sees a slightly different stereo image, resulting in a 3D effect.

2 Find out how polarisation effects are used to show the numbers on the liquid crystal display on a calculator.

ELECTROMAGNETISM

FIELDS: COULOMB'S LAW

COULOMB'S INVERSE SQUARE LAW

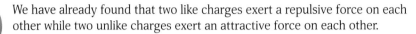

Q_1 Q_2

r

We have already found that two like charges exert a repulsive force on each other while two unlike charges exert an attractive force on each other.

Consider two point charges Q_1 and Q_2 separated by a distance r.

Coulomb found that the **force F** between the charges is **directly proportional** to the **magnitude** of each **charge** and **inversely proportional** to the **square** of the **distance** separating the charges.

$$F \propto \frac{Q_1 Q_2}{r^2}$$

$$F = \frac{1}{4\pi\varepsilon_0} \frac{Q_1 Q_2}{r^2} \quad \text{where } \frac{1}{4\pi\varepsilon_0} \text{ is the } \textbf{constant of proportionality.}$$

The constant ε_0 is called the **permittivity of free space** and has a value of $8\cdot85 \times 10^{-12} \, C^2N^{-1}m^{-2}$.

The constant of proportionality $\frac{1}{4\pi\varepsilon_0}$ has a value $9 \times 10^9 \, Nm^2C^{-2}$.

Example

The diagram shows two positive point charges separated by a distance of $0\cdot15\,m$. Calculate the magnitude and direction of the electrostatic force exerted on the $+3\cdot0\,\mu C$ charge.

$+3\cdot0\mu C$ $+2\cdot0\mu C$

$0\cdot15$ m

Solution:

$$F = \frac{1}{4\pi\varepsilon_0} \frac{Q_1 Q_2}{r^2}$$

$$= 9 \times 10^9 \frac{(3 \times 10^{-6}) \times (2 \times 10^{-6})}{0\cdot15^2}$$

$= 2\cdot4\,N$ to the left as two positive charges repel.

Note: the force on the $+2\,\mu C$ will be $2\cdot4\,N$ to the right.

DON'T FORGET ➕

The direction is found using 'like charges repel' and 'unlike charges attract'.

⚙ EXERCISE

1 Calculate the magnitude and direction of the force on the smaller point charge in (a) and (b).

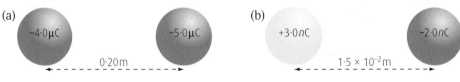

(a) $-4\cdot0\mu C$ $-5\cdot0\mu C$ (b) $+3\cdot0nC$ $-2\cdot0nC$

$0\cdot20m$ $1\cdot5 \times 10^{-2}m$

Electrostatic force between three collinear point charges

The **force on a point charge** due to the presence of two other point charges is the **vector sum** of the two forces caused by these other charges. The vector sum is straightforward if all three charges are collinear.

Example

Calculate the electrostatic force on the $-3\cdot0\,\mu C$ point charge as shown in the diagram.

$+5\mu C$ $-3\mu C$ $+4\mu C$

$0\cdot22m$ $0\cdot14m$

contd

 DON'T FORGET

Don't round answers until the end.

Solution:

Force due to $+5 \cdot 0\,\mu C = \frac{1}{4\pi\varepsilon_0}\frac{Q_1 Q_2}{r^2} = 9 \times 10^9 \frac{(5 \times 10^{-6}) \times (3 \times 10^{-6})}{0 \cdot 22^2} = 2 \cdot 79\,N$ to the left.

Force due to $+4 \cdot 0\,\mu C = \frac{1}{4\pi\varepsilon_0}\frac{Q_1 Q_2}{r^2} = 9 \times 10^9 \frac{(4 \times 10^{-6}) \times (3 \times 10^{-6})}{0 \cdot 14^2} = 5 \cdot 51\,N$ to the right.

Resultant force on $-3 \cdot 0\,\mu C = 5 \cdot 51\,N - 2 \cdot 79\,N = 2 \cdot 7\,N$ to the right.

⚙ EXERCISE

2   Calculate the electrostatic force on the $+4\,\mu C$ charge in the above diagram.

Electrostatic force between three non-collinear point charges

Finding the electrostatic force when the point charges are not in a straight line is illustrated in the following example.

Example

A point charge Q_1 is located near two other charges Q_2 and Q_3 as shown. The magnitudes of the charges are shown on the diagram. Calculate the electrostatic force on Q_1, the $+6 \cdot 5\,\mu C$ point charge.

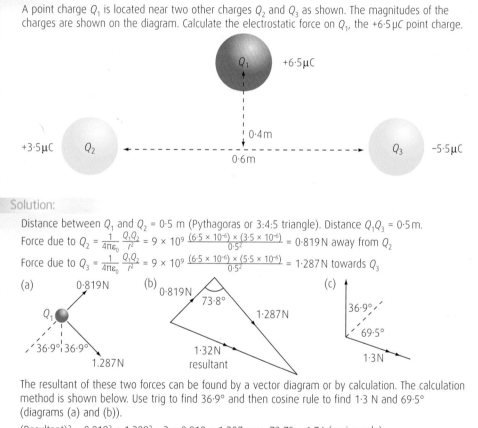

Solution:

Distance between Q_1 and $Q_2 = 0 \cdot 5$ m (Pythagoras or 3:4:5 triangle). Distance $Q_1 Q_3 = 0 \cdot 5$ m.

Force due to $Q_2 = \frac{1}{4\pi\varepsilon_0}\frac{Q_1 Q_2}{r^2} = 9 \times 10^9 \frac{(6 \cdot 5 \times 10^{-6}) \times (3 \cdot 5 \times 10^{-6})}{0 \cdot 5^2} = 0 \cdot 819\,N$ away from Q_2

Force due to $Q_3 = \frac{1}{4\pi\varepsilon_0}\frac{Q_1 Q_2}{r^2} = 9 \times 10^9 \frac{(6 \cdot 5 \times 10^{-6}) \times (5 \cdot 5 \times 10^{-6})}{0 \cdot 5^2} = 1 \cdot 287\,N$ towards Q_3

The resultant of these two forces can be found by a vector diagram or by calculation. The calculation method is shown below. Use trig to find $36 \cdot 9°$ and then cosine rule to find $1 \cdot 3$ N and $69 \cdot 5°$ (diagrams (a) and (b)).

$(Resultant)^2 = 0 \cdot 819^2 + 1 \cdot 289^2 - 2 \times 0 \cdot 819 \times 1 \cdot 287 \times \cos 73 \cdot 7° = 1 \cdot 74$ (cosine rule)

Resultant $= \sqrt{1 \cdot 74} = 1 \cdot 32\,N = 1 \cdot 3\,N$

Bearing $= 36 \cdot 87° + 69 \cdot 5° = 106 \cdot 4° = 106°$ (diagram (c))

Force on $+6 \cdot 5\,\mu C$ point charge $= 1 \cdot 3\,N$ at $106°$

⊙ ONLINE

Visit www.brightredbooks.net for an online Coulomb's Law calculator.

⚙ EXERCISE

3   Calculate the resultant force on the $-5 \cdot 5\,\mu C$ point charge in the above diagram.

💭 THINGS TO DO AND THINK ABOUT

In a 'show' question involving Coulomb's law where the numerical answer is given you must start with the formula and insert the numerical value of ε_0 to show you have substituted it correctly. There will be no marks for the answer as it is given. Marks are given for showing you know the steps and substitutions leading up to the answer. If you fail to show the substitution $\varepsilon_0 = 8 \cdot 85 \times 10^{-12}$ then marks will be lost. This applies to 'show' questions only.

✓ ONLINE TEST

Head to www.brightredbooks.net and test yourself on Coulomb's Inverse Square Law.

FIELDS: ELECTRIC FIELD STRENGTH

ELECTRIC FIELD STRENGTH: OVERVIEW

As described earlier, a **field** is a **region where forces exist**. Charged particles exert electrostatic forces on each other, so the **region around charged particles** is called an **electric field**.

Electric field strength E at a particular point is defined as the **force per unit positive charge** at that point. This means it is the **electrostatic force exerted on a point charge** of +1 C at that point and will require a **magnitude** and **direction**.

The magnitude of E can be calculated using this relationship.

$E = \frac{F}{Q}$ The unit of E is NC^{-1}

DON'T FORGET

Electric field strength is analogous to gravitational field strength $\left(g = \frac{F}{m}\right)$.

An electron moves between the cathode and anode in an electron gun. At one particular point, the electron experiences a force of $7 \cdot 2 \times 10^{-17}$ N towards the positive anode. Calculate the electric field strength at that point.

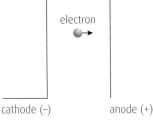

electron

cathode (−)   anode (+)

The charge on an electron is $-1 \cdot 6 \times 10^{-19}$ C.

$E = \frac{F}{Q}$

$= \frac{7 \cdot 2 \times 10^{-17}}{1 \cdot 6 \times 10^{-19}}$

$= 450 \, \text{NC}^{-1}$

The direction of the electric field at this point will be towards the cathode, as a unit positive charge will experience a force towards the negative electrode.

ELECTRIC FIELD AROUND A POINT CHARGE

Point X is a distance r from a **point charge** $+Q$. The **electric field strength** at point X will be the **force experienced by a unit positive charge** (+1 C) placed at X. Coulomb's Law is used to calculate the force between $+Q$ and +1 C a distance r apart.

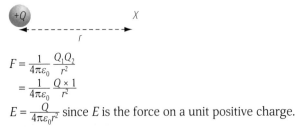

$+Q$ X

r

$F = \frac{1}{4\pi\varepsilon_0} \frac{Q_1 Q_2}{r^2}$

$= \frac{1}{4\pi\varepsilon_0} \frac{Q \times 1}{r^2}$

$E = \frac{Q}{4\pi\varepsilon_0 r^2}$ since E is the force on a unit positive charge.

⚙ EXERCISE

1 Calculate the electric field strength at
   a   30 cm from a point charge of $5 \cdot 0 \, \mu$C
   b   40 cm from a point charge of $5 \cdot 0 \, \mu$C.

2 At what distance from a $6 \cdot 0$ nC point charge will the electric field strength be 110 NC^{-1}?

DON'T FORGET

Don't confuse nC and μC.

ELECTRIC FIELD LINES

A diagram of the electric field lines around a positive point charge is shown here.

The **radial field decreases** in value as *r* increases. A **negative point charge** will have the field lines pointing **towards the negative charge**.

ELECTRIC FIELD STRENGTH BETWEEN TWO POINT CHARGES

The electric field between two point charges depends on the polarity of the charges, as shown.

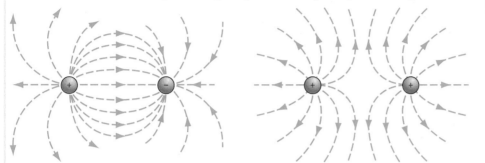

The arrows show the direction of the resultant force on a unit positive charge. The electric field strength at a particular point between the charges has a contribution from each charge. The following example shows how to calculate each contribution.

Example

Two point charges are separated by a distance of 6 cm. Calculate the electric field strength at point X.

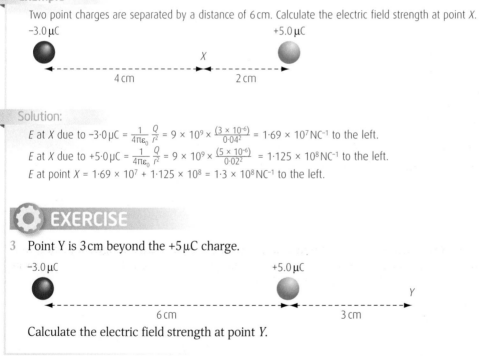

Solution:

E at X due to $-3.0\,\mu C = \frac{1}{4\pi\varepsilon_0}\frac{Q}{r^2} = 9\times10^9 \times \frac{(3\times10^{-6})}{0.04^2} = 1.69\times10^7\,NC^{-1}$ to the left.

E at X due to $+5.0\,\mu C = \frac{1}{4\pi\varepsilon_0}\frac{Q}{r^2} = 9\times10^9 \times \frac{(5\times10^{-6})}{0.02^2} = 1.125\times10^8\,NC^{-1}$ to the left.

E at point X = $1.69\times10^7 + 1.125\times10^8 = 1.3\times10^8\,NC^{-1}$ to the left.

EXERCISE

3 Point Y is 3 cm beyond the $+5\,\mu C$ charge.

Calculate the electric field strength at point Y.

THINGS TO DO AND THINK ABOUT

Electric field lines travel from positive to negative. An electron will experience a force in the opposite direction to the arrows of the field lines.

DON'T FORGET

The field lines must start on the charge surface and at right angles to the surface. Two field lines starting at the same point on the surface is incorrect.

ONLINE

Visit www.brightredbooks.net for an interactive exercise on electric field patterns.

DON'T FORGET

Field lines close together indicate a stronger electric field.

ONLINE TEST

Head to www.brightredbooks.net and test yourself on electric field strength.

FIELDS: ELECTRIC POTENTIAL

ELECTRICAL POTENTIAL: AN OVERVIEW

DON'T FORGET

Unit positive charge is +1 C

Electrical potential is analogous to gravitational potential and is another way of giving information about an electric field.

The **electrical potential at a point** in an electric field is defined as the **work done** bringing a **unit positive charge** from **infinity** to **that point**.

Integration is used to derive the relationship for the **electrical potential V** at a distance r from a point charge $+Q$.

$$V = \frac{Q}{4\pi\varepsilon_0 r}$$

The unit of V is **volt V** or JC^{-1}

The **electrical potential at infinity is zero**, as is the case with gravitational potential, but there is no negative sign in this case. This is because a repulsive force has to be overcome as a unit positive charge is moved towards $+Q$, and work must be done.

A negative point charge of $-Q$ will have negative values of V around it. Work must be done as a unit positive charge moves away from $-Q$ to eventually reach zero volts at infinity.

Example

Point X is 0·75 m from a point charge of $+5\,\mu C$. Calculate the electrical potential at point X.

Solution:

$V = \frac{Q}{4\pi\varepsilon_0 r}$

$= 9 \times 10^9 \times \frac{(5 \times 10^{-6})}{0\cdot75}$

$= 6 \times 10^4\,V$

⚙ EXERCISE

1   Calculate the electrical potential at a distance of 2 nm from an electron.

ELECTRICAL POTENTIAL INVOLVING TWO OR MORE POINT CHARGES

When two or more point charges are involved, care must be taken to insert the minus sign for negative charges and to substitute the correct distance when calculating V.

Example

Calculate the electrical potential at point X between these two point charges.

Solution:

$V = \frac{Q_1}{4\pi\varepsilon_0 r_1} + \frac{Q_2}{4\pi\varepsilon_0 r_2}$

$= 9 \times 10^9 \times \frac{(3 \times 10^{-9})}{0\cdot5} + 9 \times 10^9 \times \frac{(-4 \times 10^{-6})}{0\cdot2}$

$= 54 - 180$

$= -126\,V$

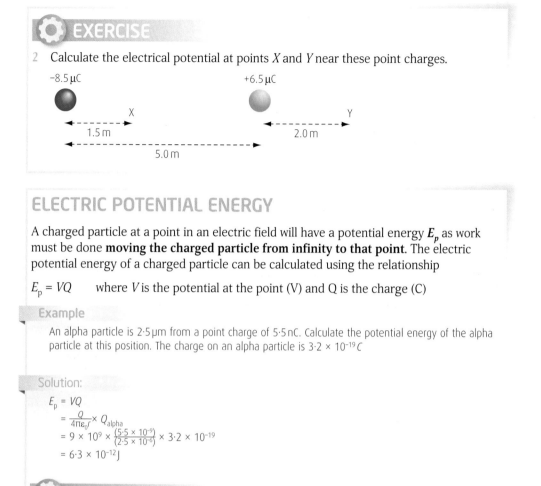

⚙ EXERCISE

2 Calculate the electrical potential at points X and Y near these point charges.

ELECTRIC POTENTIAL ENERGY

A charged particle at a point in an electric field will have a potential energy E_p as work must be done **moving the charged particle from infinity to that point**. The electric potential energy of a charged particle can be calculated using the relationship

$E_p = VQ$ where V is the potential at the point (V) and Q is the charge (C)

Example

An alpha particle is $2 \cdot 5\,\mu m$ from a point charge of $5 \cdot 5\,nC$. Calculate the potential energy of the alpha particle at this position. The charge on an alpha particle is $3 \cdot 2 \times 10^{-19}\,C$

Solution:

$E_p = VQ$

$= \frac{Q}{4\pi\varepsilon_0 r} \times Q_{alpha}$

$= 9 \times 10^9 \times \frac{(5 \cdot 5 \times 10^{-9})}{(2 \cdot 5 \times 10^{-6})} \times 3 \cdot 2 \times 10^{-19}$

$= 6 \cdot 3 \times 10^{-12}\,J$

⚙ EXERCISE

3 Calculate the electric potential energy of a proton at a distance of $3 \cdot 8 \times 10^{-10}$ m from an alpha particle.

THE ELECTRONVOLT eV

An alternative unit of energy is the electronvolt (eV). This is the energy gained (or lost) by an electron when it moves through a potential difference of one volt.

Converting one eV to joules gives $E = QV = 1 \cdot 6 \times 10^{-19} \times 1 = 1 \cdot 6 \times 10^{-19}\,J$

The electronvolt and multiples such as GeV are useful in particle physics when sub-atomic particles are accelerated in particle accelerators.

▶ VIDEO LINK

Check out the video on electric potential at www.brightredbooks.net

💭 THINGS TO DO AND THINK ABOUT

In particle physics mass and energy are often interchanged by $E = mc^2$ so the electronvolt can also be used as a unit of mass. The mass of an electron is $9 \cdot 11 \times 10^{-31}$ kg but can also be quoted as $0 \cdot 51$ MeV.

Show that $9 \cdot 11 \times 10^{-31}$ kg is consistent with $0 \cdot 51$ MeV.

Strictly speaking the unit should really be $\frac{MeV}{c^2}$ rather than MeV but scientists use either unit.

The mass of the Higgs boson was quoted by CERN as 126 GeV. Try converting this to kilograms.

Extension

This is extension work although it is a valid data handling exercise at AH level.

✓ ONLINE TEST

Head to www.brightredbooks.net and test yourself on electric potential.

FIELDS: UNIFORM ELECTRIC FIELD

UNIFORM ELECTRIC FIELD: OVERVIEW

A **uniform field** exists between **two charged parallel plates** as shown.

An expression for the electric field strength between the plates can be derived as follows.

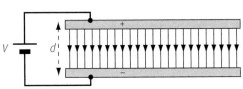

Two parallel plates have a **potential difference** V across them and are a **distance** d apart. The electric field lines have a **direction from + to −**. Work must be done moving a charge $+Q$ from the bottom plate to the top plate, as a **repulsive force** F must be overcome.

Work done = $F \times d$

An alternative expression is

Work done = QV

equating gives $F \times d = QV$

rearranging $\quad \frac{F}{Q} = \frac{V}{d}$

substituting $\quad E = \frac{F}{Q}$ gives $E = \frac{V}{d}$

⚙ EXERCISE

1 The electric field strength between two parallel plates is 2.5×10^4 NC⁻¹ when the p.d. across the plates is 2 kV. How far apart are the plates?

Example

Two parallel plates are 25 cm apart and have a potential difference of 2 kV applied across them as shown.

a Calculate the unbalanced force on an electron midway between the plates.

b The parallel plates are in a vacuum, and the electron would accelerate upwards, as there are no collisions with air molecules. Calculate
  i the electron's acceleration.
  ii the time taken to reach the top plate.

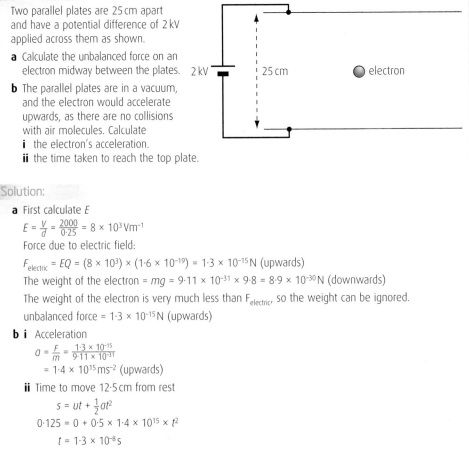

Solution:

a First calculate E

$E = \frac{V}{d} = \frac{2000}{0.25} = 8 \times 10^3$ Vm⁻¹

Force due to electric field:

$F_{electric} = EQ = (8 \times 10^3) \times (1.6 \times 10^{-19}) = 1.3 \times 10^{-15}$ N (upwards)

The weight of the electron = $mg = 9.11 \times 10^{-31} \times 9.8 = 8.9 \times 10^{-30}$ N (downwards)

The weight of the electron is very much less than F$_{electric}$, so the weight can be ignored.

unbalanced force = 1.3×10^{-15} N (upwards)

b i Acceleration

$a = \frac{F}{m} = \frac{1.3 \times 10^{-15}}{9.11 \times 10^{-31}}$

$= 1.4 \times 10^{15}$ ms⁻² (upwards)

ii Time to move 12.5 cm from rest

$s = ut + \frac{1}{2}at^2$

$0.125 = 0 + 0.5 \times 1.4 \times 10^{15} \times t^2$

$t = 1.3 \times 10^{-8}$ s

TWO-DIMENSIONAL MOTION IN A UNIFORM ELECTRIC FIELD

The previous example considered a stationary electron in a uniform electric field. Now we will consider another common situation where a charged particle has an initial speed before entering the electric field. The region between the plates is evacuated.

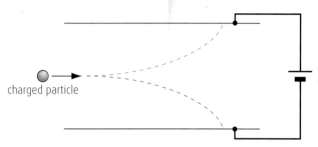

charged particle

There is no **horizontal** force on this charged particle, so it will continue to move horizontally with a steady speed (Newton's First Law).

An **electric force** will act **vertically** on the particle. If the **charge on the particle is positive**, it will experience an electric force **towards the negative plate** and will **accelerate downwards** (Newton's Second Law). The resultant parabolic trajectory of a positively charged particle is shown by the red dotted line.

A **negatively charged particle** projected between the plates would **accelerate upwards** as well as continuing with a steady speed horizontally and would follow a path similar to the blue dotted line.

The exact trajectory depends on the mass, charge and speed of the particle as well as the separation and voltage of the plates.

Example

Electrons enter an electric field midway between two deflecting plates with a speed of $2 \cdot 1 \times 10^7 \, \text{ms}^{-1}$. The plates are 80 mm long and 50 mm apart and there is a potential difference of 500 V across the plates.

Calculate

a the time an electron takes to pass the deflecting plates.

b the vertical deflection **s** of an electron as it leaves the space between the deflecting plates.

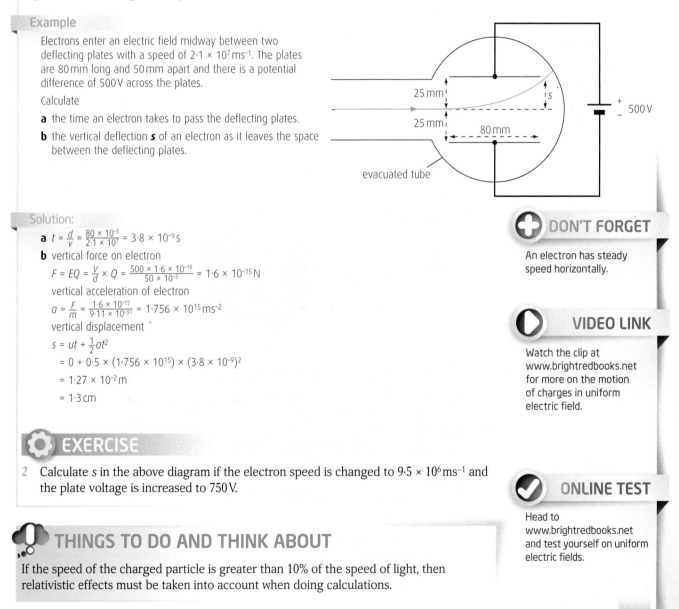

25 mm

25 mm

80 mm

s

500 V

evacuated tube

Solution:

a $t = \frac{d}{v} = \frac{80 \times 10^{-3}}{2 \cdot 1 \times 10^7} = 3 \cdot 8 \times 10^{-9} \, \text{s}$

b vertical force on electron

$F = EQ = \frac{V}{d} \times Q = \frac{500 \times 1 \cdot 6 \times 10^{-19}}{50 \times 10^{-3}} = 1 \cdot 6 \times 10^{-15} \, \text{N}$

vertical acceleration of electron

$a = \frac{F}{m} = \frac{1 \cdot 6 \times 10^{-15}}{9 \cdot 11 \times 10^{-31}} = 1 \cdot 756 \times 10^{15} \, \text{ms}^{-2}$

vertical displacement

$s = ut + \frac{1}{2}at^2$

$= 0 + 0 \cdot 5 \times (1 \cdot 756 \times 10^{15}) \times (3 \cdot 8 \times 10^{-9})^2$

$= 1 \cdot 27 \times 10^{-2} \, \text{m}$

$= 1 \cdot 3 \, \text{cm}$

⊕ DON'T FORGET

An electron has steady speed horizontally.

▶ VIDEO LINK

Watch the clip at www.brightredbooks.net for more on the motion of charges in uniform electric field.

⚙ EXERCISE

2 Calculate s in the above diagram if the electron speed is changed to $9 \cdot 5 \times 10^6 \, \text{ms}^{-1}$ and the plate voltage is increased to 750 V.

✓ ONLINE TEST

Head to www.brightredbooks.net and test yourself on uniform electric fields.

💭 THINGS TO DO AND THINK ABOUT

If the speed of the charged particle is greater than 10% of the speed of light, then relativistic effects must be taken into account when doing calculations.

FIELDS: MAGNETIC FIELDS

MAGNETIC FIELDS: OVERVIEW

We are familiar with the concept of a magnetic field from earlier studies of physics, and we will now look in some detail at the **magnetic forces that arise on conductors** and **moving charged particles in magnetic fields**. The magnetic field lines show the direction that a compass needle would be forced to align with and **not** the direction of the force on any charges.

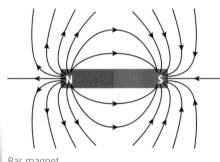

Bar magnet

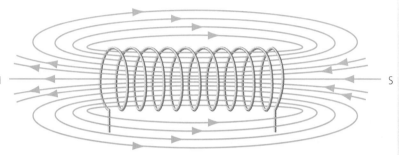

Current carrying coil

The magnetic field lines travel from **north to south** by convention. When the current in the coil is switched off, the magnetic field disappears, so the **magnetic field** must be caused by the **moving charges** (electrons) in the wire.

The Earth's magnetic field is thought to be caused by electrons moving in its molten iron core by convection currents. The Earth's magnetic field is similar to a giant bar magnet.

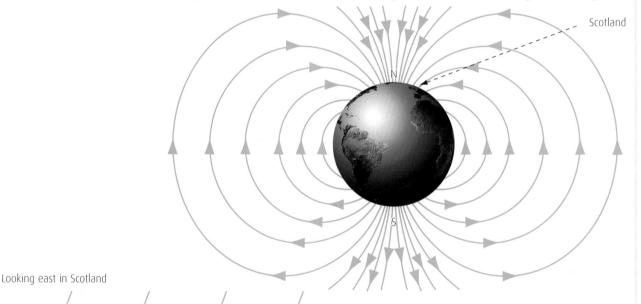

Looking east in Scotland

The Earth's magnetic field lines are horizontal at the Equator and vertical at the North Pole. In Scotland, the magnetic field lines are at an angle of approximately 69° to the horizontal. This angle increases with latitude.

Magnetic induction

The strength of a magnetic field at a point is called the **magnetic induction** and has the symbol B. The **unit of magnetic induction** is the **tesla (T)**. A more precise definition of magnetic induction B will follow.

FERROMAGNETISM

Magnetic materials like iron have groups of atoms called **domains** where many of the electrons orbiting the atoms line up with orbits parallel to each other producing a magnetic field. Domains are quite small in size but have a locally intense magnetic field. Normally in a piece of iron the domains are not lined up and the iron is unmagnetised. If the iron is placed in a magnetic field the domains line up with the magnetic field and the iron becomes magnetised.

This effect of materials becoming magnetized by an external magnetic field is called **ferromagnetism**. Iron, cobalt and nickel are ferromagnetic materials.

In bulk material the domains usually cancel, leaving the material unmagnetised

Externally applied magnetic field

The direction of an individual domain's magnetic field is shown by a small red arrow.

MAGNETIC FIELD AROUND A CONDUCTOR

The magnetic field around a long straight current-carrying conductor is a series of concentric circles centred on the conductor.

The **magnetic induction B** at a **distance r** from the conductor is **directly proportional** to the **current I** and **inversely proportional** to the **perpendicular distance r** from the conductor.

$$B \propto \frac{I}{r}$$

$$B = \frac{\mu_0}{2\pi} \frac{I}{r}$$

The constant of proportion is $\frac{\mu_0}{2\pi}$ where μ_0 is called the **permeability of free space** and has a value $4\pi \times 10^{-7}\,\text{TmA}^{-1}$. Permeability is the magnetic constant of the medium.

Example

A long straight conductor carries a current of 2·5 A. Calculate the magnetic induction at a distance of 0·050 m from the conductor.

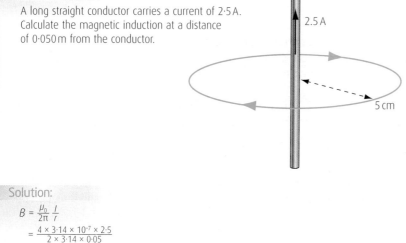

2.5 A

5 cm

Solution:

$$B = \frac{\mu_0}{2\pi} \frac{I}{r}$$
$$= \frac{4 \times 3\cdot14 \times 10^{-7} \times 2\cdot5}{2 \times 3\cdot14 \times 0\cdot05}$$
$$= 1\cdot0 \times 10^{-5}\,T = 10\,\mu T$$

VIDEO LINK

For a video link on magnetic domains visit www.brightredbooks.net

THINGS TO DO AND THINK ABOUT

1. In the above example, calculate the value of the magnetic induction at 0·20 m from the conductor.

2. What current will give a magnetic induction of 65 μT at a distance of 40 mm from a long straight conductor?

3. A long straight conductor carries a current of 8·5 A. At what distance from the conductor will the magnetic induction be 1·5 mT?

ONLINE TEST

Take the test on magnetic fields at www.brightredbooks.net

FIELDS: FORCES DUE TO MAGNETIC FIELDS

FORCE ON A CURRENT-CARRYING CONDUCTOR

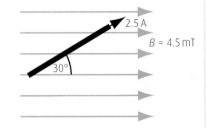

A conductor of length **L** carries a current **I** at an angle θ to a magnetic field of magnetic induction **B**. The conductor will experience a force given by the following expression.

$F = BIL\sin\theta$ (the derivation is not required)

When the conductor is perpendicular to the magnetic field, the force is

$F = BIL$ since $\theta = 90°$

When the conductor is parallel to the magnetic field, the force is zero since $\theta = 0°$.

Example

A conductor of length 60 cm carries a current in a magnetic field as shown.

Calculate the magnitude of the force exerted on the conductor.

2.5 A
B = 4.5 mT
30°

Solution:

$F = BIL\sin\theta$
$= (4.5 \times 10^{-3}) \times 2.5 \times 0.6 \times \sin30°$
$= 3.4 \times 10^{-3} \, N$

The direction of this force is **not** in the direction of **B** or the **current**. A special mnemonic called the **right-hand rule** is used to help predict the direction of this force.

Right-hand rule

The direction of the force on a current-carrying conductor in a magnetic field is:
- perpendicular to the plane carrying the conductor
- perpendicular to the direction of the magnetic field.

The right-hand rule uses the thumb and first two fingers of the right hand.

Arrange the thumb and first two fingers at right angles to each other as shown.

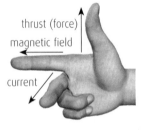

thrust (force)
magnetic field
current

⚙ EXERCISE

A straight conductor of length 40 cm is placed in two different magnetic fields carrying different currents in each situation as shown. Calculate the magnitude and direction of the force on the conductor in each case.

a

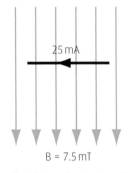

25 mA

B = 7.5 mT

b

× × × × × ×

× × × × × ×
0.35 A

× × × × × ×

× × × × × ×

× × × × × ×

B = 4.5 mT (into page)

Definition of magnetic induction

A magnetic field has a **magnetic induction** of **1 tesla** when a conductor of **length 1 metre**, carrying a **current of 1 ampere** perpendicular to the field, is acted on by a **force of 1 newton**.

contd

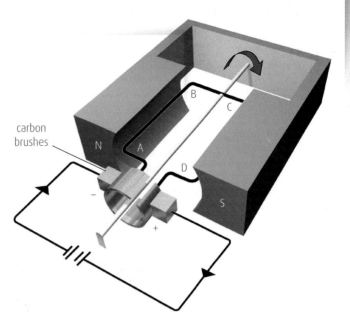

ELECTRIC MOTOR

A simple electric motor consists of a single-turn coil of wire which can rotate about an axis in a magnetic field.

The coil is initially horizontal. When a current flows through the brushes into the coil, length **AB** experiences a force F **upwards**. (Use the right-hand rule to confirm the direction.) Length **CD** experiences a force F **downwards** as the electron current flows from C to D. Side **BC** experiences **no force** as the current is parallel to the magnetic field.

The coil experiences **two torques** which cause the coil to turn clockwise, as seen from the front.

Torque $= F \times \left(\frac{1}{2}BC\right) + F \times \left(\frac{1}{2}BC\right)$

$= 2 \times \left(F \times \frac{1}{2}BC\right)$

The magnetic poles have a concave shape, so the magnetic field is radial. As the coil rotates, the **torque will be constant**, as the **force** and the **perpendicular distance** between **force** and **axis** will both be constant.

Example

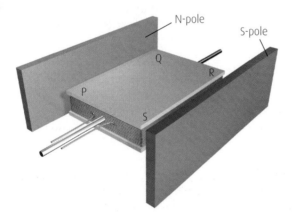

A rectangular coil *PQRS* of a model electric motor consists of 25 turns of wire with *PQ* = 60 mm and *QR* = 30 mm. The magnetic induction between the poles of the magnet is 0·15 T and the current in the coil is 1·8 A. Calculate the torque on the coil when the coil is horizontal as shown.

Solution:

Torque on side *PQ* $= (F \times d) \times 25$

$\qquad\qquad\qquad = \left(BIL \times \frac{1}{2}QR\right) \times 25$

$\qquad\qquad\qquad = 0{\cdot}15 \times 1{\cdot}8 \times (60 \times 10^{-3}) \times (15 \times 10^{-3}) \times 25$

$\qquad\qquad\qquad = 6{\cdot}075 \times 10^{-3}\,\text{Nm}$

Torque on side *RS* $= 6{\cdot}075 \times 10^{-3}\,\text{Nm}$

Torque on the coil $= 6{\cdot}075 \times 10^{-3} \times 2 = 1{\cdot}2 \times 10^{-2}\,\text{Nm}$.

This motor does not have a radial magnetic field. As the coil begins to rotate the torque will decrease as the distance between the force direction and the axis of rotation decreases.

THINGS TO DO AND THINK ABOUT

Some physics textbooks will use 'conventional current', where current direction is taken from positive to negative. These textbooks will use a different rule for the direction of the force on the conductor but will still give the same result as the right-hand rule.

ONLINE

Visit www.brightredbooks.net for an interactive simulation of an electric motor. (Don't be put off by their use of conventional current. Scientists and engineers of tomorrow have to be multi-skilled!)

ONLINE TEST

Test yourself on this topic at www.brightredbooks.net

FIELDS: THE FOUR FUNDAMENTAL FORCES

MEASURING *B* EXPERIMENTALLY

The magnetic induction **B** between two magnets can be measured experimentally using the relationship $F = BIL\sin\theta$ and a current balance.

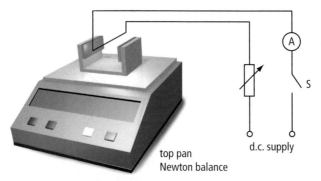

top pan
Newton balance

d.c. supply

A known length of wire is suspended rigidly between the magnets at right angles to the field. ($\theta = 90°$)

When switch **S** is closed, a **force is exerted upwards** on the wire. An **equal and opposite force** is exerted on the top pan balance **downwards**. The reading on the digital balance in **g** or **mg** is converted to **newtons**.

Typical results

$I = 200\,\text{mA}$

balance reading = 25 mg $\qquad \Rightarrow F = (25 \times 10^{-3}) \times 10^{-3} \times 9\cdot8\,\text{N}$

$L = 4\,\text{cm}$

$\theta = 90°$

$$B = \frac{F}{IL\sin\theta}$$
$$= \frac{25 \times 10^{-6} \times 9\cdot8}{200 \times 10^{-3} \times 4 \times 10^{-2} \times 1}$$
$$= 3\cdot1 \times 10^{-2}\,\text{T}$$

The magnetic field can also be measured directly by using a Hall probe or even some smartphones.

MILLIKAN'S OIL-DROP EXPERIMENT

Extension

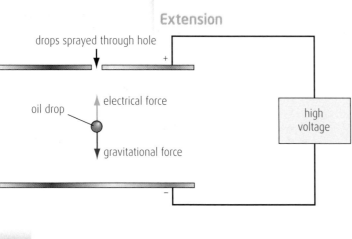

drops sprayed through hole

oil drop

electrical force

gravitational force

high voltage

Millikan's oil drop experiment is listed in the syllabus but not in the mandatory part.

In 1909, Millikan sprayed charged oil droplets into an electric field between two parallel plates. By varying the voltage **V** across the plates, he was able to stop a particular droplet from moving downwards by balancing the weight of the oil drop **mg** with an upwards electric force **EQ** or $\frac{VQ}{d}$ where **Q** is the charge on the oil droplet and **d** is the separation of the plates.

$$mg = \frac{VQ}{d}$$

contd

The mass of the oil drop was found after measuring its terminal velocity after **V** was switched off. The charge on the oil drop could be calculated when all the other variables were known. Millikan measured the charges on thousands of oil drops and noticed a pattern in the values of **Q**. He concluded that the values of electric charge were all integral multiples of **1·6 × 10⁻¹⁹ C (e)**. For example, 8·0 × 10⁻¹⁹ C = 5e and 1·92 × 10⁻¹⁸ C = 12e. A charge of 4·0 × 10⁻¹⁹ C is not possible as this would be $2\frac{1}{2}e$ and is not an integer multiple of *e*.

THE FOUR FUNDAMENTAL FORCES OF NATURE

Physicists have identified **four fundamental forces** of nature. The **gravitational force** and **electromagnetic force** are two of these forces and have been studied earlier in this book.

The other two forces are called the **strong force** and the **weak force** and these forces only exist inside the atom.

The strong force

The strong force holds all the protons and neutrons together inside the nucleus. The **strong force** overcomes the repulsive forces between positive protons and **holds the nucleus together**.

The **range** of the strong force is **very short** – less than 10⁻¹⁴ m.

As the name implies, the strong force is the strongest of the four fundamental forces.

The weak force

The **weak force** is responsible for the **emission of a beta particle** from the nucleus of a radioactive element.

A **neutron** inside the nucleus **decays** into a **proton** and an **electron** (the **beta particle**). The weak force overcomes the attractive force between the beta particle and a proton and ejects the beta particle from the nucleus. The weak force is a **short-range force** found only inside the nucleus.

Gravitational and electromagnetic forces

We have already seen that gravitational and electromagnetic forces extend over great distances.

The following example compares these two forces between the proton and orbiting electron in a hydrogen atom where the orbit radius is 5·3 × 10⁻¹¹ m.

Gravitational force	Electromagnetic force
$F = \dfrac{Gm_p m_e}{r^2}$	$F = \dfrac{1}{4\pi\varepsilon_0}\dfrac{Q_p Q_e}{r^2}$
$= \dfrac{6\cdot67 \times 10^{-11}\,(1\cdot67 \times 10^{-27}) \times (9\cdot11 \times 10^{-31})}{(5\cdot3 \times 10^{-11})^2}$	$= 9 \times 10^9 \times \dfrac{(1\cdot6 \times 10^{-19}) \times (1\cdot6 \times 10^{-19})}{(5\cdot3 \times 10^{-11})^2}$
$= 3\cdot6 \times 10^{-47}\,\text{N}$	$= 8\cdot2 \times 10^{-8}\,\text{N}$

The electromagnetic force is greater than the gravitational force by a factor of 10³⁹, so the gravitational force is negligible in this case.

THINGS TO DO AND THINK ABOUT

Take care with orders of magnitude. The electromagnetic force is 39 orders of magnitude greater than the gravitational force. This does not mean it is 39 times greater but 10³⁹ times greater.

The mass of a neutron is greater than the mass of an electron.

a How many times greater is it?

b How many orders of magnitude greater is it?

DON'T FORGET

The beta particle is an electron which originates in the nucleus. It is not an orbiting electron.

DON'T FORGET

'A factor of 10³⁹ greater' can be stated as '39 orders of magnitude greater'.

VIDEO LINK

For a video link on the four fundamental forces visit www.brightredbooks.net

ONLINE TEST

Take the test on fields at www.brightredbooks.net

CIRCUITS: CAPACITORS IN D.C. CIRCUITS

In higher physics you will have investigated the charging and discharging of capacitors. This is now studied in more detail.

TIME CONSTANT FOR CR CIRCUIT.

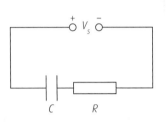

An uncharged capacitor of capacitance C farads is connected to a series resistor of resistance R ohms and a d.c. supply voltage V_S volts. The graph shows how the current in the circuit changes with time.

The current decreases exponentially from an initial value of I_0.

The current I in this circuit is related to the time t of charging by the relationship

$$I = I_0 e^{-t/RC}$$

The product RC is called the time constant τ for the circuit drawn.

The unit of τ (or RC) is the unit of time (s).

This can be shown as follows

$$R \times C = \frac{V}{I} \times \frac{Q}{V} = \frac{Q}{I} = t$$

DON'T FORGET

$I_0 = \frac{V_s}{R}$

Example

An 80 µF uncharged capacitor is connected to a 50 kΩ resistor and a supply voltage V_S. Calculate the time constant for this circuit.

Solution:

$R \times C = 50\,000 \times 80 \times 10^{-6} = 4\,s$.

What does this mean? After a time of 4 seconds (or one time constant) the charging current in the circuit can be determined.

$I = I_0 e^{-t/RC}$

$I = I_0 e^{-4/4}$

$I = I_0 e^{-1}$

$I = I_0 \times 0.368$

$I = 37\% \times I_0$

After one time constant (4 s) the charging current has fallen to 37% of the initial charging current I_0.

Alternatively the charging current has fallen by 63% of its initial value I_0 in one time constant.

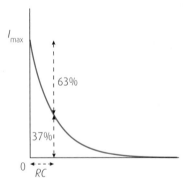

The following graph shows how the charging current changes as the time increases in units of time constant.

After two time constants the charging current has fallen to

$$I = I_0 e^{-t/RC} = I_0 e^{-2RC/RC} = I_0 e^{-2} = 0.135 I_0.$$

After five time constants, the capacitor can be considered fully charged as the charging current is now $0.007 I_0$. Use the above relationship to show that this is the case.

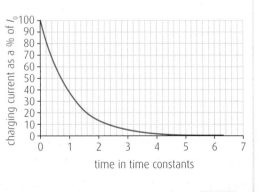

contd

Example

The graph shows how the current varies with time as a capacitor charges in a CR circuit

Determine the time constant of the capacitor.

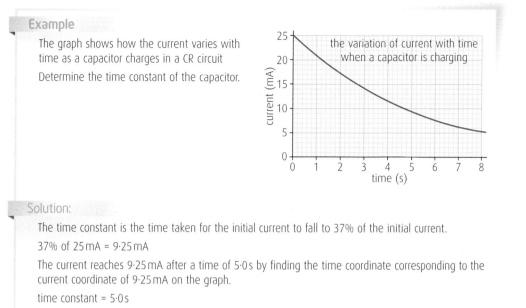

the variation of current with time when a capacitor is charging

Solution:

The time constant is the time taken for the initial current to fall to 37% of the initial current.

37% of 25 mA = 9·25 mA

The current reaches 9·25 mA after a time of 5·0 s by finding the time coordinate corresponding to the current coordinate of 9·25 mA on the graph.

time constant = 5·0 s

If the resistance in the CR circuit is 15 kΩ then the capacitance can be calculated.

CR = time constant

$C = \frac{5\cdot0}{15\,000} = 330\,\mu F$

VOLTAGE ACROSS A CAPACITOR IN A CR CIRCUIT

Consider again the CR circuit on the opposite page which shows a capacitor charging after connecting to a d.c. supply of voltage V_S.

The voltage across the capacitor (V_C) increases as the capacitor charges and the graph of V_C against time is shown in the graph.

The voltage across the charging capacitor V_C after time t is given by the relationship

$V_C = V_S (1 - e^{-t/RC})$

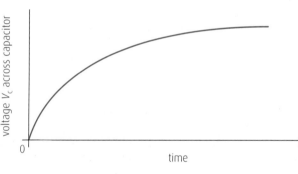

The time constant can also be calculated from the graph as it will be the time taken for the graph to reach 63% of the maximum value. Numerical values will be needed on both axes to do this.

CHARGE Q ON A CHARGING CAPACITOR IN A CR CIRCUIT

The charge Q stored on a capacitor increases as the capacitor charges. The graph of Q against time will be the same shape as the graph above. The relationship for Q is similar to the relationship above for V_C.

$Q = Q_{max} (1 - e^{-t/RC})$ where $Q_{max} = C \times V_S$

The time constant can be determined directly from the graph if values are shown on each axis. The time constant is the time taken for the charge to reach 63% of the maximum value.

 # THINGS TO DO AND THINK ABOUT

The capacitor C in the CR circuit at the top of the opposite page now discharges. The voltage V_C across this discharging capacitor will vary with time as shown by the following graph.

Suggest a (mathematical) relationship showing how the value of V_C varies with time as the capacitor discharges in the CR circuit. Explain your reasoning.

VIDEO LINK

Check out the clip at www.brightredbooks.net for time constant in RC circuit.

ONLINE TEST

Head to www.brightredbooks.net and test your knowledge of capacitors in d.c. circuits.

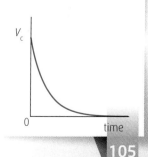

CIRCUITS: CAPACITIVE REACTANCE

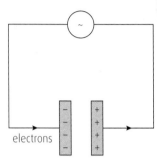

electrons

When the electrons move anticlockwise in this circuit, a negative charge builds up on the left-hand plate of the capacitor. This negative charge opposes the flow of more negative electrons onto the plate. When the a.c. current reverses, the right-hand plate becomes negative and opposes the flow of electrons on to it.

DON'T FORGET ⊕

An a.c. ammeter and a.c. voltmeter will record r.m.s. values.

DON'T FORGET ⊕

Each piece of data used in the calculation has 2 s.f. so the answer should have 2 s.f. as well.

CAPACITIVE REACTANCE: OVERVIEW

A capacitor in a circuit will oppose the flow of alternating current (a.c.) in the circuit and the term **capacitive reactance** is used to describe this opposition.

The capacitor's **opposition to a.c.** is called **capacitive reactance** and has the symbol X_C.

The unit of X_C is ohms (Ω) but capacitive reactance X_C is not the same as resistance R.

No energy is dissipated (turned into heat energy) by capacitive reactance in a capacitor unlike resistance where electrical energy is converted into heat energy in a resistor.

During the charging cycle of a capacitor, energy is supplied by the power supply moving electrons to one of the capacitor plates. During the discharge cycle, electrons are returned from the capacitor plate to the power supply with no loss of energy.

The capacitive reactance of a capacitor opposes the flow of a.c. current and a variation of Ohm's law applies.

In this circuit the power supply provides an alternating voltage with a fixed frequency f.

The capacitive reactance X_C can be determined by recording the value on the voltmeter (V_{rms}) and recording the corresponding value on the ammeter (I_{rms}) then using the relationship

$$X_C = \frac{V_{rms}}{I_{rms}}$$

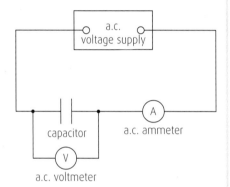

Example

The ammeter in the above circuit records a value of 0·29 A when the voltmeter records a value of 12 V. Calculate the capacitive reactance of the capacitor.

Solution:

$$X_C = \frac{V_{rms}}{I_{rms}}$$
$$= \frac{12}{0·29}$$
$$= 41\,\Omega$$

DETERMINING THE VALUE OF X_C OF A CAPACITOR EXPERIMENTALLY

The circuit shown in the previous example would use a lab power supply with variable a.c. voltage. Vary the value of the power supply voltage and record the corresponding currents on the ammeter. A table like this would be completed.

V_{rms} (V)				
I_{rms} (A)				

Plot a graph of V_{rms} against I_{rms}

The graph should be a straight line through the origin.

Find the gradient of the graph using $\frac{V_2 - V_1}{I_2 - I_1}$

X_C = gradient

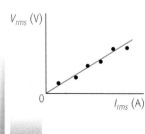

Finding the value of X_C using this method assumes the frequency of the a.c. is constant.

We will now see that X_C depends on the value the a.c. frequency.

contd

Alternative formula for capacitive reactance X_C

The alternating current I_C in a capacitive circuit is directly proportional to the a.c. frequency f.

At high frequencies the a.c. current is correspondingly high, so the capacitive reactance must be low. Similarly, the capacitive reactance must be higher at lower a.c. frequencies where the current is low.

It can be shown that $X_C \propto \frac{1}{f}$ where f is the frequency of the alternating current

and $\quad X_C = \frac{1}{C}$ where C is the capacitance of the capacitor

giving $\quad X_C = \frac{1}{2\pi f C}$

UNIT OF $\frac{1}{2\pi f C}$

The unit of capacitive reactance can be confirmed as the ohm.

The unit of $fC = s^{-1} \times \frac{Q}{V} = \frac{1}{t} \times \frac{Q}{V} = \frac{Q}{t} \times \frac{1}{V} = I \times \frac{1}{V} = \frac{I}{V}$

Inverting, we get the unit of $\frac{1}{2\pi f C} = \frac{V}{I} = \Omega$

DON'T FORGET

π is a number with no units.

Example

A capacitor of capacitance $56\,\mu F$ is connected to a 12 V a.c. lab power supply as shown.

a Calculate the capacitive reactance of the capacitor in this circuit.

b Calculate the r.m.s. current in the circuit.

12 V rms
50 Hz

$56\,\mu F$

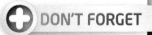

DON'T FORGET

A lab power supply is a step-down transformer and will have mains frequency (50 Hz) when the power supply is operating on a.c.

Solution:

a $X_C = \frac{1}{2\pi f C}$

$\quad = \frac{1}{2 \times 314 \times 50 \times 56 \times 10^{-6}}$

$\quad = 57\,\Omega$

b $I_{rms} = \frac{V_{rms}}{X_C}$

$\quad = \frac{12}{57}$

$\quad = 0{\cdot}21\,A$

DON'T FORGET

$56{\cdot}841\,\Omega$ would not be acceptable as it has too many s.f. for the data used.

GRAPHS INVOLVING CAPACITIVE REACTANCE X_C

X_C is inversely proportional to frequency f and directly proportional to $\frac{1}{f}$ if the capacitance C is constant.

$X_C = \frac{1}{2\pi f C}$

$X_C = \frac{1}{2\pi C} \times \frac{1}{f}$

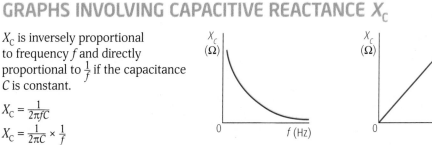

DON'T FORGET

Care must be taken when sketching a graph showing inverse proportion. The graph should **not** have an intercept on the vertical axis. And the graph line should **not** finish up parallel to either axis.

The gradient of the straight line graph X_C against $\frac{1}{f}$ is equal to $\frac{1}{2\pi C}$

If a graph of X_C against $\frac{1}{f}$ can be drawn for a capacitor of unknown capacitance, then the capacitance can be found by calculating the gradient of the straight line.

gradient $= \frac{1}{2\pi C}$

$\qquad C = \frac{1}{2\pi \times (gradient\ of\ straight\ line\ graph)}$

VIDEO LINK

For a worked example on calculating X_C visit www.brightredbooks.net

THINGS TO DO AND THINK ABOUT

The graph shows how the reactance of a capacitor varies with $\frac{1}{frequency}$.

Show that the capacitance of the capacitor is $4{\cdot}8 \times 10^{-3}\,F$.

ONLINE TEST

Head to www.brightredbooks.net and test your knowledge of capacitive resistance.

CIRCUITS: INDUCTORS IN D.C. CIRCUITS 1

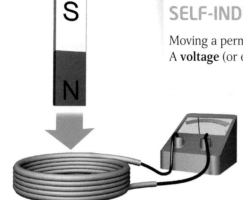

SELF-INDUCTANCE OF A COIL

Moving a permanent magnet inside a coil of wire causes a reading on the ammeter. A **voltage** (or e.m.f.) is **induced** in the coil.

Lenz's Law states that the **induced e.m.f. opposes** the **change in current** causing it.

In the diagram, an N-pole is moved into the coil. Lenz's law predicts that the current induced in the coil will produce an N-pole at the top end to repel the permanent magnet N-pole.

When the magnet is moved up **out of the coil**, the coil will produce an **S-pole** to **oppose the change** and **attract** the permanent magnet downwards.

Any **changing magnetic field** will **induce** an electromotive force (e.m.f.) in a **conductor** placed in the **magnetic field**.

Another example of a changing magnetic field is in the region around a current-carrying conductor **when the current is changing**. An increasing current will cause the magnetic field at a particular point to increase in value.

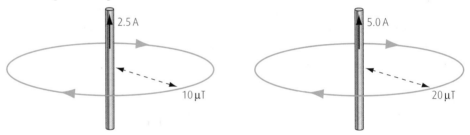

The magnetic induction **B** has a value of $10\,\mu T$ at a distance of $5\,cm$ from a conductor carrying a current of $2.5\,A$.

If the current increases to $5.0\,A$, the value of **B** at the same position will increase to $20\,\mu T$.

The magnetic field at all points around the conductor will increase as the current increases.

But the conductor itself is in its own changing magnetic field, so a **self-induced e.m.f.** will be produced across the conductor. This will be **in addition** to the potential difference causing the current in the first place.

Changing currents and changing magnetic fields will occur
- in d.c. circuits when switching the circuit on or off
- in a.c. circuits all the time.

INDUCTORS

A coil of wire has a greater magnetic field around it than a straight length of wire carrying the same current. The self-induced e.m.f. generated by a coil when its magnetic field changes can be quite considerable.

A coil of wire is often called an **inductor** because of its ability to induce an e.m.f. across itself.

Symbols for an inductor are shown here.

Air-cored inductor   Iron-cored inductor

INDUCTOR IN A D.C. CIRCUIT

Connecting an inductor into a series d.c. circuit illustrates well the effects of the self-induced e.m.f. The inductor is assumed to have negligible resistance.

Switch **S** is closed for several seconds then opened again. The graph shows how the current changes during this time.

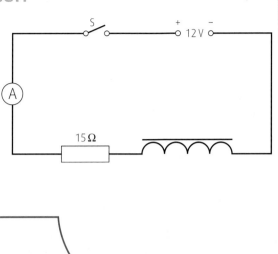

The current does not reach its maximum value immediately. This is because the increasing current causes a changing magnetic field around the inductor. An e.m.f. is induced in the conductor, and this opposes the build-up of current. The induced e.m.f. is often called a **back e.m.f.**, as it **opposes the change in current**. The **back e.m.f. reduces to zero** as the **current approaches its maximum constant value**.

The maximum current is found by Ohm's law.

$$I = \frac{V}{R} = \frac{12}{15} = 0\cdot8\,\text{A}$$

There is a magnetic field around the inductor, but it is not changing, so the back e.m.f. is zero at maximum current.

On opening the switch, the current reduces, and this causes a changing (collapsing) magnetic field. A back e.m.f. is induced in the inductor, opposing the decreasing current. The decreasing current lasts for a short time after the switch is opened.

An inductor with **more turns** will create a **bigger back e.m.f.** and the current will take **longer** to reach its maximum value.

An inductor with **fewer turns** will create a **smaller back e.m.f.** and the current will take **less time** to reach its maximum value.

Notice that the **maximum current will be unchanged** if the inductors have negligible resistance.

(A more exact way of defining the size of an inductor will follow on the next page.)

Lenz's Law and conservation of energy

If the induced e.m.f. did not oppose the build-up of current but instead acted the opposite way, then the current increase would be even greater, so producing a bigger induced e.m.f.. This constantly increasing cycle would create its own energy once started. The conservation of energy does not allow this, so the induced e.m.f. in an inductor must oppose the changing current.

THINGS TO DO AND THINK ABOUT

Any explanation of self-inductance must include details about the changing magnetic field as this causes a voltage to be induced.

DON'T FORGET

A datalogging device connected across the inductor can display the graph directly on the screen of a laptop.

VIDEO LINK

For a video demonstration of Lenz's law visit www.brightredbooks.net

ONLINE TEST

Head to www.brightredbooks.net and test your knowledge of indicators in d.c. circuits.

CIRCUITS: INDUCTORS IN D.C. CIRCUITS 2

INDUCTANCE

The previous page discussed how a self-induced e.m.f. is produced in a coil of wire whenever the current in the coil changes. We also saw evidence that the self-induced e.m.f. opposes the change in current. Mathematically, the relationship is direct proportion.

$\mathcal{E} \propto -\frac{dI}{dt}$ where $\mathcal{E}$ is the **self-induced e.m.f.** in a coil or inductor.

The **negative sign** shows that $\mathcal{E}$ is in the **opposite direction to** $\frac{dI}{dt}$.

Introducing the **constant of proportion**, we get

$\mathcal{E} = -L\frac{dI}{dt}$ where L is the **self-inductance** or **inductance** of the coil.

The unit of L is $\frac{V}{As^{-1}} = VsA^{-1} = \mathbf{H}$ (henry)

A coil has an **inductance of 1 henry** when an **e.m.f. of 1 volt** is **self-induced** when the **current changes** at a rate of **1 ampere per second**.

An inductor with an inductance of 2·5 H will induce a back e.m.f. of 2·5 V when the rate of change of current is 1 As⁻¹.

Example

Calculate the self-induced e.m.f. in an inductor of inductance 0·75 H when the rate of change of current is 3·5 As⁻¹.

Solution:

$\mathcal{E} = -L\frac{dI}{dt}$

$= -0.75 \times 3.5$

$= -2.6\,V$

Example

A back e.m.f. of 8 V is induced in a coil when the current changes at 2 As⁻¹. Calculate the inductance of the coil.

Solution:

$\mathcal{E} = -L\frac{dI}{dt}$

$-8 = -L \times 2$ or $L = -\frac{\mathcal{E}}{\frac{dI}{dt}} = -\frac{(-8)}{2} = 4\,H$

$L = 4\,H$

ENERGY STORED IN AN INDUCTOR

An inductor **stores energy** in the magnetic field around it. The relationship for the energy E stored in an inductor is

$E = \frac{1}{2}LI^2$ where L is the **inductance** of the inductor and I is the **current** in the inductor.

 **EXERCISE**

1   Calculate the energy stored in a 4·0 H inductor carrying a current of 600 mA.

2   A 50 mH inductor stores 160 μJ of energy. Calculate the current flowing in the inductor.

contd

DON'T FORGET

Some teachers encourage students to substitute numerical values into a formula at the start before rearranging the formula. This guarantees the substitution mark and any subsequent mistake rearranging is treated as an arithmetic slip with the minimum penalty.

DON'T FORGET

$\mathcal{E}$ has a **negative** sign. The minus sign must be included when substituting for $\mathcal{E}$ or back e.m.f.

DON'T FORGET

It is incorrect physics to simply drop the minus sign to get a positive value for L.

DON'T FORGET

Remember to **square the current** during calculations for energy – a common oversight.

Example

This is an example of an exam-type question of an inductor in a d.c. circuit. The circuit shows an inductor of negligible resistance and a battery of negligible internal resistance.

The graph shows the growth in the current after switch S is closed.

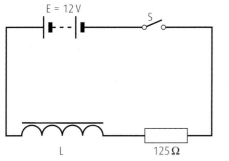

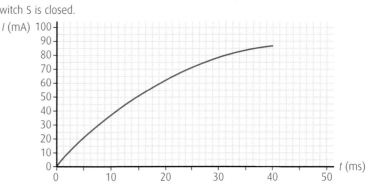

a What is the magnitude of the back e.m.f. at the instant that the switch is closed?

b Use the graph to calculate the initial rate of change of current.

c Calculate the inductance of the coil.

d Calculate the back e.m.f. 25 ms after switch **S** is closed.

e Calculate the maximum energy stored in the inductor.

Solution:

a back e.m.f. = 12 V since $I = 0$; back e.m.f. is equal and opposite to the battery e.m.f.

b At $t = 0$, $\frac{dI}{dt}$ = gradient of graph at the origin

$= \frac{45 \times 10^{-3} - 0}{10 \times 10^{-3} - 0}$ = using points (0,0) and (10 ms, 45 mA)

$= 4 \cdot 5 \, \text{As}^{-1}$ (Note: depending on tangent drawn, $\frac{dI}{dt}$ could be in the range 4–5 As^{-1}.)

c $\mathcal{E} = -L\frac{dI}{dt}$

$-12 = -L \times 4 \cdot 5$ or $L = -\frac{\mathcal{E}}{\frac{dI}{dt}} = -\frac{(-12)}{4 \cdot 5} = 2 \cdot 7 \, \text{H}$

$L = 2 \cdot 7 \, \text{H}$

d $\quad\quad\quad\quad\quad\quad t = 25 \, \text{ms} \Rightarrow I = 72 \, \text{mA}$

V across 125 Ω resistor $= IR = 72 \times 10^{-3} \times 125$

$= 9 \, \text{V}$

$\therefore$ back e.m.f. $= 12 \, \text{V} - 9 \, \text{V}$

$= 3 \, \text{V}$

e $I_{max} = \frac{V}{R} = \frac{12}{125}$

$= 9 \cdot 6 \times 10^{-2} \, \text{A}$

$E = \frac{1}{2}LI^2$

$= 0 \cdot 5 \times 2 \cdot 7 \times (9 \cdot 6 \times 10^{-2})^2$

$= 1 \cdot 2 \times 10^{-2} \, \text{J}$

 **DON'T FORGET**

In this equation, $\mathcal{E}$ must have a minus sign.

ONLINE

For a video on inductance visit www.brightredbooks.net

ONLINE TEST

Head to www.brightredbooks.net and test your knowledge of inductors in d.c. circuits.

THINGS TO DO AND THINK ABOUT

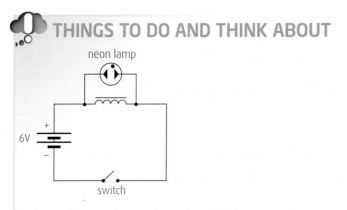

Lighting the neon lamp takes at least 70 V to ionise the neon gas inside the lamp.

Explain how a 6 V battery is able to make the lamp light in this circuit.

CIRCUITS: INDUCTORS IN A.C. CIRCUITS

INDUCTORS IN A.C. CIRCUITS

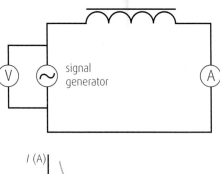

signal generator

In an **a.c. circuit**, the **current** and its associated **magnetic field** are continually changing. An inductor in an a.c. circuit will **always self-induce a back e.m.f.** and not just when switching on or off as was the case in a d.c. circuit.

As the **frequency** of the a.c. **increases**, the **rate of change of current will increase** and we would expect the **back e.m.f.** across an inductor to **increase**. The following experiment looks at how the current varies as the frequency is increased in an inductive circuit.

A signal generator is used as the a.c. power supply. This allows the frequency of the alternating current to be varied, and the frequency can be read from the dial on the signal generator. It is important to keep the output voltage of the supply constant, so an a.c. voltmeter is connected across the output terminals of the signal generator. An a.c. ammeter takes readings of the r.m.s. current in the circuit for various frequencies set by the signal generator, and a graph of **current I** against **frequency f** is drawn.

I (A) ... *f (Hz)*

The graph shows an inverse-type relationship between I and f. As the **frequency increases**, the r.m.s. **current** in the circuit **decreases**.

f	I	f × I

To investigate further what type of relationship exists between **current I** and **frequency f**, we should look for some combination of the two variables which gives a constant. For inverse relationships, multiply the variables together. Record this in an extra column in the results table.

If $f \times I$ is a constant, then

$$f \times I = k$$

$$I = \frac{k}{f}$$

$I \propto \frac{1}{f}$ **k** is constant of proportionality.

Alternatively, plot a graph of **I** against $\frac{1}{f}$ and look for a straight line through the origin as proof of $I \propto \frac{1}{f}$.

I (A) ... *$\frac{1}{f}$ (Hz^{-1})*

The graphical method is more time-consuming than the evaluation of the constant of proportionality.

Inductive reactance

The opposition of an inductor to alternating current is called **inductive reactance, X_L ohms**.

X_L is proportional to the a.c. frequency, so the **current decreases** as the **frequency increases**.

$$X_L \propto f \quad \text{also} \quad X_L \propto L$$

Combining leads to the relationship

$$X_L = 2\pi f L$$

Show that the unit of X_L (or $2\pi f L$) is the ohm (Ω).

Example

A 3·5 mH inductor is part of a circuit connected to the mains electricity supply. Calculate the inductive reactance of the inductor in this circuit.

Solution:

$X_L = 2\pi f L$

$= 2 \times 3{\cdot}14 \times 50 \times (3{\cdot}5 \times 10^{-3})$

$= 1{\cdot}1\,\Omega$

contd

There is a second formula for inductive reactance X_L involving V_{rms} and I_{rms}.

$X_L = \dfrac{V_{rms}}{I_{rms}}$ where V_{rms} is the r.m.s. voltage across the inductor

and I_{rms} is the r.m.s current in the inductor.

Some numerical exam questions may require both $\dfrac{V_{rms}}{I_{rms}}$ and $X_L = 2\pi fL$ being used in the solution. The following worked example shows an extended exam question involving both formulae.

Example

An inductor of inductance 5·8 H and negligible resistance is connected to this circuit.

The signal generator has an output voltage of 12 V.

The reading on the ammeter is 3·8 mA.

Calculate the output frequency of the signal generator.

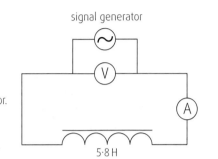

signal generator

5·8 H

Solution:

The inductive reactance X_L of the inductor must be found first.

$X_L = \dfrac{V_{rms}}{I_{rms}}$

$\quad = \dfrac{12}{3 \cdot 7 \times 10^{-3}}$

$\quad = 3243 \, \Omega$

$X_L = 2\pi fL$

$3243 = 2 \times 3 \cdot 14 \times f \times 5 \cdot 8$

$\quad f = 89 \, Hz$

Uses of inductors

There are many uses of inductors in both d.c. and a.c circuits.

A coil in an a.c. circuit will filter out unwanted high frequency electrical signals (interference). The coil (or inductor) will have a large value of inductive reactance X_L when the frequency is high. The high frequency interference becomes negligible as a result.

Traffic-light cameras

Traffic-light speed cameras are used in conjunction with two induction loops buried under the road junction. A metal car above a buried coil will act like an iron core and increase the inductance of the coil. If this happens in the first coil and then the second coil when the lights are at red, a picture is taken.

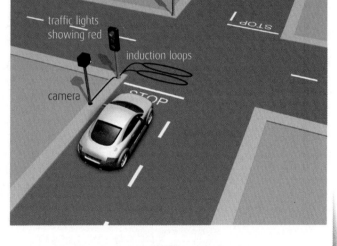

traffic lights showing red

induction loops

camera

An inductor is a coil of wire and will inevitably have a resistance as well as an inductive reactance. The combined effect of **resistance** and **reactance** is called **impedance** with the symbol Z and unit ohm.

Modern headphones have a typical impedance of 20 Ω.

Extension

Impedance is not in the AH syllabus but could well feature in an exam if the physics of how to combine resistance and reactance is given.

ELECTROMAGNETIC RADIATION

UNIFICATION OF ELECTRICITY AND MAGNETISM

In 1865, the Scottish physicist James Clerk Maxwell succeeded in linking electricity and magnetism in his theory of electromagnetic radiation.

Prior to this both electricity and magnetism were treated as separate branches of physics although the links between them had been established experimentally.

In 1820, Hans Christian Øersted had shown that electric currents exerted forces on a magnetic compass.

In 1831, Michael Faraday noted that changing magnetic fields could induce electric currents.

Maxwell postulated that an oscillating electric charge produced an oscillating electric field. This oscillating electric field induced an oscillating magnetic field which in turn induced an oscillating electric field and so on. He reasoned that these oscillating fields must be in the form of a self-propagating travelling wave and he produced mathematical relationships to support his theory. Maxwell called this wave an **electromagnetic wave**.

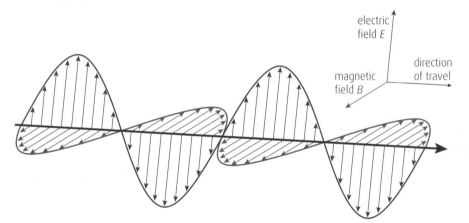

The diagram shows a polarised electromagnetic wave travelling to the right. The electric field, magnetic field and the direction of travel of the wave are all **perpendicular** to each other.

An unpolarised electromagnetic wave would have many (red) electric field components oscillating in many planes travelling to the right like a 3D sausage. Each electric field component would be in phase with a corresponding magnetic field component (in blue) and the diagram of an unpolarised electromagnetic wave showing all the electric and magnetic field components could be difficult to follow.

SPEED OF LIGHT

Maxwell published a set of partial differential equations which describe how electric and magnetic fields are generated and interact with each other. These equations included the permittivity of free space ε_0 and the permeability of free space μ_0. Maxwell's equations predicted that the speed of electromagnetic waves is given by the relationship

$$speed = \frac{1}{\sqrt{\varepsilon_0 \mu_0}}$$

Substituting the values of ε_0 and μ_0 into this equation gives

$$speed = \frac{1}{\sqrt{8 \cdot 85 \times 10^{-12} \times 4\pi \times 10^{-7}}}$$

$$= 3 \cdot 00 \times 10^8 \, m\,s^{-1}$$

contd

As this value is the same as the speed of light Maxwell deduced that light must be an electromagnetic wave.

 EXERCISE

The permittivity and permeability of free space are measured experimentally in order to find a value for the speed of light.

The experimental results are

permittivity of free space = $8.8 \times 10^{-12} \pm 9 \times 10^{-13}\,C^2N^{-1}m^{-2}$

permeability of free space = $9.9 \times 10^{-7} \pm 1.5 \times 10^{-7}\,TmA^{-1}$

a   Show that these results give a value for the speed of light of $3.4 \times 10^8\,ms^{-1}$.

b   Show that the result for the permeability of free space has the largest percentage uncertainty.

c   Show that the percentage uncertainty in the value for the speed of light is $\pm 9\%$

d   Show that the absolute uncertainty in the value for the speed of light is $\pm 3 \times 10^7\,ms^{-1}$.

e   Comment on the accuracy of the experimental results

f   Show that the units of ε_0 and μ_0 are consistent with the unit of speed in the relationship $c = \frac{1}{\sqrt{\varepsilon_0\mu_0}}$

RADIO WAVES

Maxwell also predicted the existence of a whole family of electromagnetic waves, each with different frequencies. By the 1870s only three types of electromagnetic waves had been identified: visible light, infrared waves and ultraviolet light. In 1887, Heinrich Hertz produced radio waves thus confirming Maxwell's earlier theory.

Sadly Maxwell had died in 1879, aged just 48, and did not witness this proof of his earlier theoretical work.

 ONLINE

Visit www.brightredbooks.net for a video on Maxwell's four laws.

THINGS TO DO AND THINK ABOUT

James Clerk Maxwell is without doubt Scotland's greatest physicist and has been described as one of the top three physicists of all time alongside Newton and Einstein.

Maxwell's equations, however, are quite mathematically complex and are normally studied at university rather than at school. This could be the reason why he is less well known by the wider public.

He is credited with the world's first colour photograph taken as part of a lecture in London.

And his theory that Saturn's rings must be made of small particles was confirmed in the 1980s by the Voyager spacecraft during a fly-by.

Maxwell's achievements are certainly worth some research.

 **ONLINE TEST**

Head to www.brightredbooks.net and test yourself on electromagnetic radiation.

Statue of James Clerk Maxwell on George Street, Edinburgh.

UNCERTAINTIES

MEASUREMENT UNCERTAINTIES 1

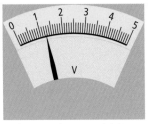

Analogue scale – uncertainty $\pm \frac{1}{2}$ of the smallest division

Digital scale - uncertainty ± 1 of the least significant digit

MEASUREMENT UNCERTAINTIES

All measurements of physical quantities are subject to uncertainties. These uncertainties can be expressed in **absolute** or **percentage** terms, such as:

$d = 2.5 \pm 0.5$ mm ± 0.5 mm is the **absolute** uncertainty

$d = 2.5 \pm 20\%$ mm   0.5 is 20% of 2.5 mm

Reading uncertainty

The **reading uncertainty** is a measure of how accurately an instrument's **scale** can be read.

The reading uncertainty on an analogue scale is $\pm \frac{1}{2}$ **of the smallest division**. The reading uncertainty on a digital scale is **± 1 of the least significant digit**.

In the diagrams, the voltmeter reading is 1.25 ± 0.05 V. The thermometer reading is 24.2 ± 0.1°C.

Random uncertainty

Repeating a measurement n times usually gives a **spread** of measurements about a **mean**.

For example, if repeated measurement of the time for an event gives times of 3.2, 3.6, 3.3, 3.2, 3.7 seconds

the mean time is $= \frac{\Sigma\ measurements}{n}$ Random uncertainty $= \frac{max\ measurement - min\ measurement}{n}$

$$= \frac{17.0}{5} \qquad\qquad\qquad\qquad\quad = \frac{3.7 - 3.2}{5}$$

$$= 3.4\,s \qquad\qquad\qquad\qquad\quad\; = \frac{0.5}{5}$$

$$\qquad\qquad\qquad\qquad\qquad\qquad\qquad\; = 0.1\,s$$

time $= 3.4 \pm 0.1$ s

Systematic uncertainty

A **systematic uncertainty** is an error which affects all the measurements in the same way, making them all either too high or too low, such as forgetting to zero a digital balance before a series of measurements.

Calibration uncertainty

Calibration uncertainties are given by manufacturers of scientific instruments as an indication of the accuracy of these instruments.

Typical calibration uncertainties in common lab instruments are shown in the table.

wooden metre stick	± 0·5 mm
steel ruler	± 0·1 mm
thermometer (liquid in glass)	± 0·5°C
analogue meter	± 2% of full scale deflection
digital meter	± 0·5% of the reading + 1 in the least significant digit

Example

What is the calibration uncertainty of a digital voltmeter with a reading of 2·58 V?

Solution:

Calibration uncertainty = ± 0·5% of the reading + 1 in the least significant digit

$$= \pm\ 0.5\% \times 2.58 + 0.01$$

$$= \pm\ 0.013 + 0.01$$

$$= \pm\ 0.023\,V$$

contd

Total uncertainty for a measurement

Each measurement in physics should have both a **reading** and a **calibration** uncertainty. If the measurement is repeated several times, there will also be a **random** uncertainty. Each of these three uncertainties contributes to an overall or **total uncertainty** for the measurement.

total uncertainty = $\sqrt{(reading\ uncert)^2 + (calibration\ uncert)^2 + (random\ uncert)^2}$

Example

The time of 10 oscillations of a pendulum is measured 5 times using a digital stopwatch.
The results in seconds are
17·9, 18·4, 17·4, 18·2, 17·6
Calculate the total uncertainty.

Solution:

Mean time = $\frac{\Sigma\ measurements}{n} = \frac{89·5}{5} = 17·9\,s$

Random uncertainty = $\frac{max\ measurement - min\ measurement}{n} = \frac{18·4 - 17·4}{5} = \frac{1·0}{5} = \pm\,0·2\,s$

Reading uncertainty = $\pm\,0·1\,s$

Calibration uncertainty = ± 0·5% of the reading + 1 in the least significant digit

$\qquad\qquad = \pm\,0·5\% \times 17·9 + 0·1$

$\qquad\qquad = \pm\,0·09 + 0·1$

$\qquad\qquad = \pm\,0·19\,s$

Total uncertainty in time measurement

$\qquad\qquad = \pm\,\sqrt{(reading\ uncert)^2 + (calibration\ uncert)^2 + (random\ uncert)^2}$

$\qquad\qquad = \pm\,\sqrt{(0·1)^2 + (0·19)^2 + (0·2)^2}$

$\qquad\qquad = \pm\,\sqrt{0·086}$

$\qquad\qquad = \pm\,0·29$

$\qquad\qquad = \pm\,0·3\,s$

time of 10 oscillations = 17·9 ± 0·3 s

Dominant uncertainty

Uncertainties less than $\frac{1}{3}$ of the **dominant uncertainty** can be ignored. This often simplifies the procedure to find the uncertainty associated with a measurement.

Example

The uncertainties in the measurement of a temperature are recorded as
· reading uncertainty = ± 0·5°C   · calibration uncertainty = ± 0·5°C   · random uncertainty = ± 0·1°C.
Calculate the total uncertainty.

Solution:

The random uncertainty is less than $\frac{1}{3}$ of the other two uncertainties so can be ignored.

total uncertainty in temperature = $\pm\sqrt{(reading\ uncert)^2 + (calibration\ uncert)^2}$

$\qquad\qquad = \pm\sqrt{(0·5)^2 + (0·3)^2}$

$\qquad\qquad = \pm\,0·7°C$

Taking several measurements of the temperature will reduce the random uncertainty, and in this example it can be ignored compared to reading and calibration uncertainties. Taking **even more** temperature measurements will have **no effect** on the total uncertainty in temperature, as the other uncertainties are dominant **in this example**.

ONLINE

Visit www.brightredbooks.net for more details on treatment of uncertainties in Advanced Higher Physics.

THINGS TO DO AND THINK ABOUT

A current is measured using a digital ammeter. Calculate the total uncertainty of the current measurement with this data:
- reading uncertainty = ± 0·1 mA
- calibration uncertainty = ± 0·11 mA
- random uncertainty = ± 0·4 mA.

ONLINE TEST

Test yourself on measurement uncertainties online at www.brightredbooks.net

MEASUREMENT UNCERTAINTIES 2

UNCERTAINTY IN A PRODUCT OR QUOTIENT OF QUANTITIES

Consider the relationships $X = Y \times Z$ or $X = \frac{Y}{Z}$. The percentage uncertainty in X $(\%\Delta X)$ is found using

$$\%\Delta X = \pm \sqrt{(\% \text{ uncert in } Y)^2 + (\% \text{ uncert in } Z)^2}$$

Example

The moment of inertia of an object about an axis and its angular velocity about the same axis are

$I = 9{\cdot}9 \times 10^{-4} \pm 4 \times 10^{-5}\,\text{kgm}^2$

$\omega = 4{\cdot}3 \pm 0{\cdot}3\,\text{rads}^{-1}$

Calculate the angular momentum of the object and the absolute uncertainty in the calculated value.

Solution:

$L = I\omega = (9{\cdot}9 \times 10^{-4}) \times 4{\cdot}3 = 4{\cdot}3 \times 10^{-3}\,\text{kgm}^2\text{s}^{-1}$

% uncertainty in $I = \frac{4 \times 10^{-5}}{9{\cdot}9 \times 10^{-4}} \times 100 = 4\%$

% uncertainty in $\omega = \frac{0{\cdot}3}{4{\cdot}3} \times 100 = 7\%$

% uncertainty in $L = \pm\sqrt{(\% \text{ uncert in } I)^2 + (\% \text{ uncert in } \omega)^2}$

$\qquad\qquad = \pm\sqrt{(4)^2 + (7)^2}$

$\qquad\qquad = \pm 8\%$

8% of $4{\cdot}3 \times 10^{-3} = 3 \times 10^{-4}$

$\qquad\qquad L = 4{\cdot}3 \times 10^{-3} \pm 3 \times 10^{-4}\,\text{kgm}^2\text{s}^{-1}$

⚙ EXERCISE

1 Two parallel plates are separated by a distance d and have a voltage V across them.

Measurements of the distance and voltage are

$V = 1250 \pm 100\,\text{V}$

$d = 2{\cdot}50 \times 10^{-2} \pm 1 \times 10^{-3}\,\text{m}$

Calculate the electric field strength E between the plates and the associated percentage and absolute uncertainties in E.

If more than two quantities are combined by multiplication or division, then the formula is extended.

$$\%\Delta W = \pm \sqrt{(\% \text{ uncert in } X)^2 + (\% \text{ uncert in } Y)^2 + (\% \text{ uncert in } Z)^2 + \ldots}$$

UNCERTAINTY IN A QUANTITY RAISED TO A POWER

If $P = A^n \Rightarrow$ % uncertainty in $P = n \times$ % uncertainty in A

Example

$\omega = 51 \pm 2\,\text{rads}^{-1}$. Calculate ω^2 and the uncertainty in ω^2.

Solution:

$\omega^2 = 51^2 = 2{\cdot}6 \times 10^3\,\text{rad}^2\text{s}^{-2}$

% uncertainty in $\omega = \frac{2}{51} \times 100 = 4\%$

% uncertainty in $\omega^2 = 2 \times 4 = 8\%$

$\omega^2 = 2{\cdot}6 \times 10^3 \pm 8\%$

$\omega^2 = 2{\cdot}6 \times 10^3 \pm 2 \times 10^2\,\text{rad}^2\text{s}^{-2}$

UNCERTAINTY IN A SUM OR DIFFERENCE OF QUANTITIES

The uncertainty in a sum or difference of quantities uses absolute uncertainties and not percentage uncertainties.

If $X = Y + Z$, or $X = Y - Z$

absolute $\Delta X = \pm \sqrt{(\text{absolute } \Delta Y)^2 + (\text{absolute } \Delta Z)^2}$

Example

Use this data to calculate the value and uncertainty in $v + v_s$ (used in Doppler effect)

$v = 340 \pm 5\,\text{ms}^{-1}$

$v_s = 30 \pm 3\,\text{ms}^{-1}$

Solution:

$v + v_s = 340 + 30 = 370\,\text{ms}^{-1}$

uncertainty in $v + v_s = \pm \sqrt{(3^2 + 5^2)} = \pm 6\,\text{ms}^{-1}$

$v + v_s = 370 \pm 6\,\text{ms}^{-1}$

ONLINE

Follow the link at www.brightredbooks.net for an uncertainty calculator.

THINGS TO DO AND THINK ABOUT

1   Calculate the energy stored in an inductor and the uncertainty in energy using this information:

$L = 2.5 \pm 0.2\,\text{H}$

$I = 8.3 \pm 0.5\,\text{mA}$

2   An analyser in a polarisation experiment is rotated relative to a fixed protractor.

ONLINE TEST

Test yourself on measurement uncertainties online at www.brightredbooks.net

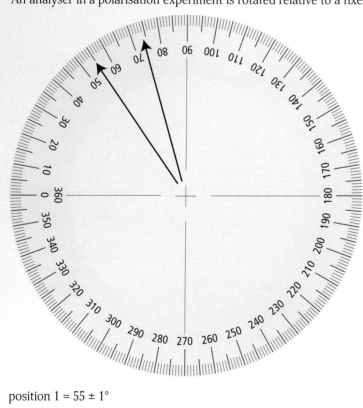

position 1 = $55 \pm 1°$

position 2 = $74 \pm 1°$

Calculate the angle of rotation and its uncertainty.

GRAPHICAL ANALYSIS

OVERVIEW

Uncertainty in the position of a point on a graph is shown using **error bars**. The lengths of the horizontal and vertical bars is a measure of the uncertainty in each coordinate.

The table of results from a current balance experiment contains uncertainties, and the corresponding graph illustrates how error bars are drawn.

Current/mA	0	100 ± 10	200 ± 10	300 ± 10	400 ± 10	500 ± 10	600 ± 10
Reading on balance/mg	0 ± 1	11 ± 1	25 ± 2	35 ± 2	48 ± 2	58 ± 3	75 ± 3

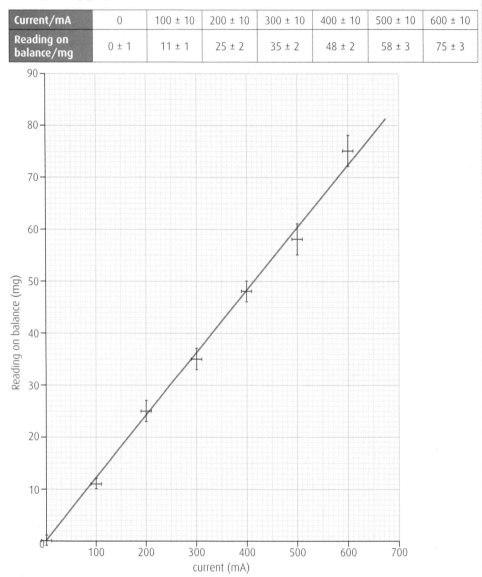

Each point is drawn with its error bars.

For straight-line graphs, the best straight line should pass through at least one error bar from each plotted point.

Straight-line graphs should also pass through the centroid (or average x and y point). The centroid of this graph is (300, 36).

Note that the uncertainties in the balance reading increase, so the sizes of the balance-reading error bars increase to reflect this. The uncertainty in the current stays constant at ± 10 mA so the horizontal (current) error bars are all the same length.

The graph relates to the experiment on page 102 to find the value of B using the formula $F = BIL\sin\theta$ where $F = mg$ and $\theta = 90°$.

$mg = (BL) \times I$

so for a graph of mg against I, the gradient $= BL$ and $B = \frac{gradient}{L}$

The gradient of the straight line graph can be calculated manually or it can be found using the LINEST function of Excel. Enter the table of results into a spreadsheet like Excel and the gradient and absolute uncertainty in the gradient of the best straight line will be calculated for you if you carry out a few simple steps (see video link).

DON'T FORGET

$mg = (BL)I$ is like $y = mx$

VIDEO LINK

For an excellent tutorial on using LINEST visit www.brightredbooks.net

⚙ EXERCISE

Using LINEST show that the gradient of the line is 0·122143 and the absolute uncertainty in the gradient is ± 0·003242.

Finding B

gradient = 0·122143 ± 0·003242 (NA^{-1}) (± 2·65%)   from LINEST

$L = 40 ± 2$ mm (± 5%)   (typical result)

$B = \frac{gradient}{L} = \frac{0.122143}{40 \times 10^{-3}} = 3.05\,T$

% uncert in $B = ± \sqrt{(\% \ uncert \ in \ gradient)^2 + (\% \ uncert \ in \ L)^2}$

$= \sqrt{2.65^2 + 5^2}$

$= 5.66\%$

$B = 3.05\,T ± 5.66\%$

$= 3.05 ± 0.17\,T$

💭 THINGS TO DO AND THINK ABOUT

ONLINE TEST

Head to www.brightredbooks.net to test yourself on uncertainties.

Practice using the LINEST function of Excel in these examples.

1   The following measurements were made of the velocity of a vehicle at time intervals of 2 seconds.

velocity (ms⁻¹)	0	0·70	1·7	2·1	3·2	4·0	4·5
time (s)	0	2·0	4·0	6·0	8·0	10	12

Enter the results into Excel.
a   Produce a graph of velocity against time. The graph should have minor gridlines.
b   Determine the acceleration of the vehicle from the information displayed.
c   Determine the absolute uncertainty in this acceleration.

DON'T FORGET

The acceleration will be the gradient of your graph.

2   The period T of a simple pendulum was measured for various lengths L. The results are shown in the table.

L (m)	0·20	0·40	0·60	0·80	1·0	1·2
T (s)	0·84	1·3	1·5	1·8	2·0	2·2
T² (s²)						

DON'T FORGET

You can use Excel to calculate T^2 for you.

The period T and length L are linked by the relationship

$T = 2\pi \sqrt{\frac{L}{g}}$ where g is the acceleration due to gravity.

a   Complete the row showing T^2.
b   Enter the results of L and T^2 into Excel
c   Plot a graph of T^2 against L using LINEST.
d   Determine the acceleration due to gravity from the information displayed.
e   Determine the uncertainty in the value for g.

PROJECT REPORT

An in-depth investigation into a physics topic must be undertaken as part of the Advanced Higher Physics course. The investigation or project should involve three or four related experiments. You are required to produce a report after completing your Advanced Higher project and the report will be submitted to SQA for marking. The project report is worth 30 marks out of a total of 130 marks for the whole course.

The report will be marked under six categories and SQA guidance on completing the report and how it is marked should be provided by your school. Some **additional** advice in the preparation of a project report is included here under the six categories.

> **DON'T FORGET** ⊕
>
> There is no need to underline headings. A change of font size and/ or emboldening looks better than underlining headings.

ABSTRACT (1 MARK)

A brief abstract or summary should follow the contents page and state the aims and findings of the project. Use the heading 'Abstract' and keep it separate from the Introduction. Many students fail to include the findings in this part of the report. For example, the aims might be to *measure the refractive index of water by several different methods*, but if the results for each method are not given, the mark will not be given.

INTRODUCTION (4 MARKS)

The underlying physics should be included in the introduction. You will be carrying out physics experiments and the theory behind these experiments should be developed. Simply stating the physics formulae you will be using and what each symbol means will only score 1 mark. You should show how each formula is derived using clear diagrams. A complicated diagram can be copied and pasted from the internet if required. If you do this, you should include a reference to the source.

If you intend to draw a straight line graph during an experiment you should show how the formula is rearranged and highlight the gradient and its significance. For example, in a simple pendulum experiment to find g we use

$$T = 2\pi \sqrt{\frac{L}{g}}$$

which by squaring both sides gives

$$T^2 = \frac{4\pi^2 L}{g}$$

Plotting T^2 against L should give a straight line of gradient $\frac{4\pi^2}{g}$. So $g = \frac{4\pi^2}{gradient}$.

Developing the theory demonstrates good understanding of the physics involved.

Simply copying large chunks of theory from a reference source does not necessarily demonstrate understanding and may be considered as plagiarism.

PROCEDURES (7 MARKS)

Diagrams and/or descriptions of the apparatus (2 marks)

There should be a diagram of the apparatus used in each experiment. Unlabelled or unclear diagrams will lose marks. Photographs of apparatus can be used provided they are clear and also labelled. Circuit diagrams with component values should be shown where necessary.

Experimental procedure (2 marks)

A clear description of how the apparatus was used in each experiment is required. The report should be written in the past tense impersonal style. For example:

contd

The radius of the ball bearing was measured using a digital micrometer

rather than

I measured the radius of the ball bearing using a micrometer.

Failure to use the past tense impersonal style will lose one of the two marks for descriptions.

Appropriate level for AH physics (3 marks)

The experiments you carry out should be at AH physics level of complexity and not experiments from N5 or Higher. Finding the acceleration due to gravity using a trolley running down a slope would not normally be considered AH level. Discuss with your teacher which experiments you intend to do to make sure they are at the appropriate level. Three or four related experiments should be attempted.

RESULTS (8 MARKS)

Data sufficient and relevant (1 mark)

Show all the readings and measurements made during your experimental work.

Analysis of data (4 marks)

The project report must show how the data was used to calculate numerical values. Sample calculations are required. The use of graphical analysis to find a physics constant is appropriate at AH level. Simply substituting numbers into a formula to get a physics constant is not normally considered AH level. Use of spreadsheet packages to produce graphs is to be encouraged. However, these graphs should not be too small and minor gridline must be included on both axes.

Uncertainties (3 marks)

Each measurement you make should have scale reading, random and calibration uncertainties where appropriate and these should be combined for a total measurement uncertainty. You only need to give one sample calculation of the total uncertainty of a measurement.

A physics constant which is calculated from a formula should have an absolute or percentage uncertainty found by combining the other uncertainties in the formula

The absolute uncertainty in the final result should have **one significant figure only** e.g.

$$g = 9{\cdot}6 \pm 0{\cdot}5\,\text{ms}^{-2} \qquad \text{or} \qquad g = 9{\cdot}73 \pm 4 \times 10^{-2}\,\text{ms}^{-2}$$

DISCUSSION (8 MARKS)

Conclusion (1 mark)

Your report should include overall conclusion(s) which should tie in with the aims stated in the abstract. The conclusion should be valid for the experimental results obtained.

Evaluation of experimental procedures (3 marks)

Include an evaluation after each experiment discussing the accuracy and precision of your measurements, sources of uncertainty, limitations of equipment and control of variables.

Avoid the temptation to say that 'better' results would be obtained by 'more accurate equipment'.

One of your graphs may be a straight line but just misses the origin. Comment on this and discuss which systematic uncertainty may be responsible.

An experimental result for the acceleration due to gravity of $g = 9{\cdot}4 \pm 0{\cdot}2\,\text{ms}^{-2}$ should have an attempt at explaining why the result is less than the accepted value and why the uncertainty limits do not extend to the accepted value.

ONLINE

For more information about managing your Advanced Higher project report, head to www.brightredbooks.net

contd

Evaluation of the investigation as a whole (3 marks)

Have this as a separate heading at the end of your report. Avoid simply restating discussion points from the individual evaluations. Describe something that went wrong and how it was resolved. Think of improvements and further work that could be done.

Quality of the project report (1 mark)

This mark is awarded by the examiner for the quality of the report. A report which has cut corners by having very little underlying physics, no sample calculations or only a few sentences for evaluation is unlikely to achieve this mark.

PRESENTATION (2 MARKS)

Structure (1 mark)

The report should have an appropriate **title**, **contents page** and **page numbers**. Having 'AH Project' as a title would not be appropriate. Missing the occasional page number by inserting a late graph will not penalised.

References (1 mark)

At least three references must be cited correctly in the main body of your report and the same ones also listed correctly at the back of the report under a heading **References**. The references must be in a standard form like the Vancouver style of referencing. Here a superscript number, often in square brackets [1], is placed in the text of the report with the numbers appearing consecutively on the References page with more detail:

> author   title   edition   place of publication
>
> 1. Tyler F, A laboratory manual of physics, 2nd ed., London, Edward Arnold, 1959, pp22–23
>
>   publisher   year of publication   page number(s)

Web pages should have as much relevant detail as possible including the URL and date accessed:

> 2. SQA, AH physics project-report general assessment information, http://www.sqa.org.uk/files_ccc/GAInfoAHPhysics.pdf visited 20 June 2015

Some references should appear in the Introduction part of your report citing the source of your physics material.

The project report should be between 2000 and 3000 words in length excluding the title page, contents page, tables, graphs, diagrams, calculations, references, acknowledgements and any appendices. The word count should be submitted with the project report. If the word count exceeds the maximum by 10%, a penalty will be applied.

DON'T FORGET

An unnumbered graph page between P16 and P17 can be numbered by hand as 16A.

THINGS TO DO AND THINK ABOUT

There are other standard forms of referencing. A Bibliography system lists the references alphabetically. This may be more time-consuming than the Vancouver system which lists references in numerical order of appearing in the text.

The Harvard system of referencing places the italicised name of the author and title in the text in brackets (*Tyler, A laboratory manual of physics*) rather than Arabic numerals [1].

Scientific and medical reports generally use the Vancouver system of referencing.

ACCURACY AND PRECISION

OVERVIEW

Non-scientists often think the terms 'accuracy' and 'precision' mean the same thing when speaking about experimental measurements and results. These terms however have quite separate meanings in science.

Measurements are said to be **accurate** when their mean value is **equal to** or **close to** the **true value**.

Measurements are said to be **precise** when they all give the **same or similar values** when repeated. The random uncertainty is small as the results are close together.

Different combinations of accuracy and precision are possible. Measurements can be
- accurate and precise
- accurate but not precise
- precise but not accurate
- neither accurate nor precise.

These different combinations can be illustrated by looking at the results of rifle shots at four separate targets.

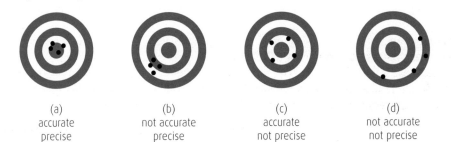

(a)	(b)	(c)	(d)
accurate	not accurate	accurate	not accurate
precise	precise	not precise	not precise

Target (b) suggests that there is a systematic error with the sights on this rifle.

Target (c) can be described as accurate as the mean shot position is on the bull's eye.

The accuracy and precision of some experimental results can be discussed when these results are graphed. For example, in your AH project you may plot a graph similar to inductive reactance X_L against frequency f. This graph should be a straight line through the origin. Your graph may look like this, however.

In your project report evaluation you can discuss the possibility of a systematic error being present and the results could be described (with reasons) as being precise but not accurate.

DON'T FORGET

This graph is a straight line but misses the origin.

THINGS TO DO AND THINK ABOUT

Sketch 'direct proportion' graphs which could be described as
- accurate but not precise
- not accurate and not precise
- accurate and precise.

Write a sentence or two after each graph explaining your choice. Accuracy and precision of measurements and results should be one of the discussion points in the evaluation sections of your project report.

OPEN-ENDED QUESTIONS

OVERVIEW

Open-ended questions are designed to encourage creative, critical thinking and written language skills appropriate to AH physics. There are many ways of answering an open-ended question in physics but there should be some physics formula(e) or principle discussed in your answer.

The AH exam will contain two open-ended questions each worth 3 marks. They can be identified by containing the expression 'Use your knowledge of physics to ...'.

The marking scheme for open-ended questions is common across all levels and subjects.

0 marks – the answer demonstrates **no understanding** (of the physics involved).

1 mark – the answer demonstrates **limited understanding**.

2 marks – the answer demonstrates **reasonable understanding**.

3 marks – the answer demonstrates **good understanding**.

Good understanding does not require a perfect answer and it can have an occasional error.

Practising open-ended questions is recommended. The following two examples may help.

DON'T FORGET

The space provided for your answer can take a maximum of 20 lines of writing.

DON'T FORGET

Alexander's Dark Band can show up well in photographs and many artists over the centuries have captured it in paintings of rainbows.

ONLINE

Visit www.brightredbooks.net to see some strategies involving physics for answering this question.

EXERCISE

1   The slightly darker region between a primary and secondary rainbow is called **Alexander's Dark Band** after Alexander of Aphrodisias who first described it in 200 AD.

A student makes the following statement about Alexander's Dark Band: 'This must be an interference effect as there is a dark region between two bright regions.'

Use your knowledge of physics to comment on the student's statement.

2   The Earth is the only planet in the Solar System with only one moon.

A physics class discusses the scenario where the Earth has more than one moon.

The class concludes that this would have very little effect on our way of life.

Use your knowledge of physics to discuss this conclusion.

THINGS TO DO AND THINK ABOUT

The following factual information **may** be useful when preparing for open-ended questions: typical mass/weight/height of an adult; typical distances (m and km); converting distances (ly to m or m to ly); typical car speed (mph and ms^{-1}); walking/running speeds; various temperatures; audio frequencies. This list is in addition to **your knowledge of physics**.

INDEX